D1636163

Health Hazards
of VDTs?

WILEY SERIES IN INFORMATION PROCESSING

Consulting Editor
Mrs Steve Shirley OBE, *F. International Limited, UK*

Visual Display Terminals
A. Cakir, D. J. Hart, and T. F. M. Stewart

Managing Systems Development
J. S. Keen

Face to File Communication
Bruce Christie

APL—A Design handbook for Commercial Systems
Adrian Smith

Office Automation
Andrew Doswell

Health Hazards of VDTs?
B. G. Pearce

Health Hazards of VDTs?

Edited by

B. G. Pearce

The HUSAT Research Group
The University of Technology
Loughborough

JOHN WILEY & SONS

Chichester · New York · Brisbane · Toronto · Singapore

Library of Congress Cataloging in Publication Data:

Main entry under title:
Health Hazards of VDTs?
   (Wileys series in information processing)
   Based on the proceedings of three one-day meetings held in
Loughborough in 1980 and 1981 sponsored by the HUSAT
(Human Sciences and Advanced Technology) Research Group.
   Includes index.
   1. Video display terminals—Hygienic aspects—
Congresses. I. Pearce, B. G. II. HUSAT Research Group
(Loughborough University of Technology) III. Title: Health
Hazards of V.D.T.s? IV. Series.
RC965.V53H4   1984   621.3819'532   82-21841
ISBN 0 471 90065 6

British Library Cataloguing in Publication Data:

Health Hazards of VDTs?—(Wiley series in information
processing)
   1. Information display systems—Congresses
   2. Industrial hygiene—Congresses
   I. Title   II. Pearce, B.G.
   363.1'89   TK7882.T6

ISBN 0 471 90065 6

Phototypeset by Millford Reprographics Int. Ltd., Luton
and printed by Page Brothers, Norwich

Contents

Section 2

Section 3

Preface

With the growing number of visual display units (VDUs)* appearing in offices there has been a corresponding increase in concern as to whether they constitute a health hazard to their users. There have been allegations ranging from minor transitory problems to more severe, irreversible effects. It is obviously of major concern to many people to assess whether any credence should be given to these allegations and, if so, whether these problems arise directly from all VDUs, from certain kinds of VDUs in certain circumstances, or from circumstances unrelated to the VDU.

The HUSAT (Human Sciences and Advanced Technology) Research Group undertakes, as part of its role, the collection and dissemination, to all interested parties, of knowledge about human issues associated with the use of information technology. Accordingly in 1980 and 1981, three public, one-day meetings were held in Loughborough under the general title of 'Health Hazards of VDUs?' The '?' in the title must be emphasized; we hoped these meetings would present the state of the art and reveal whether there were health hazards and, if so, what form they took, what factors led to them, and what actions could be taken to avoid them. This book is based on these meetings.

The book records the contributions of some of the leading authorities to the emotive debate on the safety, health, and quality of working life of the users of 'new technology'. The leading authorities include not only those scientists from a wide range of disciplines who were invited to present papers but also the representatives of equipment suppliers, users, medical advisers, and trade unionists who attended the meetings. The contributions range from reports of well controlled laboratory studies to anecdotal comments clearly influenced by vested interest.

The contributions have been grouped into three sections that reflect the content of the three one-day meetings. The first examined the evidence for the alleged direct health hazards that might affect the safety of VDU users. The second, subtitled 'Some Solutions to the Common Problems', considered the indirect effects of VDU work that might affect the health and comfort of users. The third, subtitled 'A Positive Approach to the Future', considered some of the various ways by which the quality of working life of VDU users might be improved. Each paper was followed by a discussion session and each meeting ended with a general discussion. This book includes a summary of these discussions. In the majority of the discussions the questioner has not been

*The terms Visual Display Unit (VDU) and Visual Display Terminal (VDT) are used interchangeably throughout this book. The former is in more common use in the UK; the latter is predominant in Europe and the USA.

identified. Where a contribution to the discussion is attributed to a named individual, attempts were made to check that the transcript contained a faithful interpretation of their contribution. I am grateful to all contributors, both speakers and discussants, for enabling this book to provide a comprehensive record of this many-sided and, in places, controversial debate.

I hope that this book will enable people to approach the question of the possible health hazards of working with VDUs in a more informed way. Without knowledge and understanding of the facts and the opinions held by both the informed and ill-informed, there is a danger of over-dramatization and ill-founded rumours spreading, or alternatively, of ignorance allowing malpractice to flourish.

The book does not present an agreed, well documented answer to the question 'Health Hazards of VDTs?'. It does reveal the need not only for more systematic research into the problems but also for the wider dissemination and application of our existing knowledge. It is to be hoped that the contributions will serve to provoke the necessary research and to stimulate the wider application of the known solutions to some of the problems.

B. G. PEARCE
HUSAT Research Group

Acknowledgements

The smooth running of the meetings and the production of this book are due to the combined efforts of many people whose help I would like to acknowledge. I would especially like to thank Sue Sher, for her administrative assistance before and during the meetings; Margaret Shotton, for her painstaking transcription of the tape recordings of the meetings, her translation of the spoken English into sense and for compiling the index; and Noelle Fletcher, for the many hours she spent in front of a VDU, with apparently no ill effects, processing the words.

The editor gratefully acknowledges permission from the following organizations to reproduce material in this book as follows:

Figure 1, p.62, Figure 2, p.62, Figure 3, p.63, and Figure 4, p.63, are from Fellman *et al.*, (1982) *Behaviour and Information Technology* **1**, 1. Figure 5, p.65, is from Hünting *et al.* (1981), in *Ergonomics* **24**, 12. Table 1, p.82, is from Läubli *et al.*, 'Visual impairments in VDU operations related to environmental conditions', in E. Grandjean and E. Vigliani (eds) (1980), *Ergonomic Aspects of Visual Display Terminals*. The paper by Wilhelm Hünting, p.119, was previously published in *Ergonomics* **24**, 12 (1981). Figures 1 to 8, pp.160–167, are from G. W. Radl, 'Experimental investigations for optimal presentation mode and colours of symbols on the CRT screen', in Grandjean and Vigliani (eds) (1980). Reproduced by permission of Taylor & Francis Ltd. Figure 1, p.86, is from P. T. Stone, in *The Ophthalmic Optician* **20**, 1 (1980). Reproduced by permission of *The Ophthalmic Optician*. Figure 3, p.176, Figure 4, p.177, and Figure 5, p.178, are from L. Hawkins (1982) in *Building Services and the Environmental Engineer* **4**, 8 (April 1982). Reproduced by permission of Batiste Publications Ltd. Figure 7, p.225, is reproduced, with permission, from *WRU Occasional Paper 16*, published by the Department of Employment's Work Research Unit. It is subject to Crown Copyright restrictions.

Section 1

Section 1

Chapter 1

Introduction

LEELA DAMODARAN

The HUSAT Research Group
Loughborough University of Technology, Loughborough, UK

The contributions in Section 1 fall into two major categories; those concerned with scientific investigation of potential health hazards associated with VDUs and those concerned with the wide ranging issues relating to human aspects of advanced technology. The contributors have in common a concern that the existence of any ill effects of technology should be clearly identified and thoroughly researched, to avoid undue fear and alarm among VDU users and all involved in the implementation of new technology. Section 1 focuses upon certain specific aspects of advanced technology which pose threats to the effective utilization of screen-based systems. Relevant experts in medicine, occupational hygiene, and ergonomics will present their findings and the results of the reported work will be interpreted to assist us in three ways: first, to clarify the existing state of knowledge, second, to place in perspective the reality of the health hazards associated with visual display units, and third, to identify areas requiring further research.

Whatever the nature of the 'verdict' (if any) arrived at there is one foregone conclusion: that is that all of us who are concerned with human aspects of advanced technology will have a major task ahead to disseminate the information, to give reassurance, to provide education, to develop policies and procedures for any action that is required. I emphasize this point because should we suggest that the fears we are addressing are, in fact, groundless or relate to only a small proportion of the population, the most undesirable possible outcome would be a conclusion that no action is expected or is required of any of the professional bodies involved. A state of inaction would be wholly inappropriate, if not negligent, because whatever the verdict concerning the validity of the fears, it will not in itself change the nature of the human response to new technology.

3

The relationship between people and advanced technology is complex and multi-faceted. Many will be familiar with the work of the HUSAT research group, which for more than ten years has been exploring many aspects of that relationship. For those concerned in any capacity with improving that relationship, to focus their attention exclusively upon any one hazard is to fail to grapple with the real issues. The 'reality' of the human relationship with new technology encompasses political, economic, social, psychological, and physiological attributes among others. Fears about new technology may be expressed as health hazards because these constitute the only legitimate reasons in our society to reject the technology. The only mechanisms that exist which allow people to protest about new technology are those generally associated with health hazards. Therefore to take complaints and fears about health only at their face value is to ignore the perhaps unspoken doubts and anxieties about technology de-skilling or indeed replacing the job, inabilities to cope with new procedures, a generalized loss of control over events and circumstances in this technological world. These powerful and compelling concerns will not be mollified to any significant degree by our verdict on whether VDUs really threaten their users with dermatitis or with permanent damage to the eye or whatever.

Only when the combined expertise of the many disciplines represented here begins to address not just the problem of avoiding or curing disease but also the need to positively promote the well-being of users of new technology and those whom technology has or will replace can we believe that we are moving towards a society which is in control of its technology.

The content of Section 1 is vitally important to avoid causing undue anxiety to those who use VDUs in their daily work and to ensure that every possible preventive measure is adopted where any likelihood, however small, of deleterious effects exists. But prevention of damage through the use of new technology should be only one part of our endeavours. To promote and to improve the relationship between people and technology at every level must be our overriding concern. To this end Sections 2 and 3 are concerned with positively promoting well-being through the application of known techniques, of known procedures, and of advanced technology. The difficulties are great and the challenges at all levels, not least at the methodological level, are enormous but until we can begin effectively to meet these challenges we should not delude ourselves into believing that we are making 'Progress'. We hope that we are moving forward from simply examining the hazards which might exist to looking for mechanisms and ways of resolving them. The contributors will be concerned to indicate the ways in which their organizations are already attempting to proceed and some of the problems which they have faced.

Chapter 2

Health Hazards in Perspective

B. G. PEARCE

The HUSAT Research Group
Loughborough University of Technology, Loughborough, UK

The HUSAT research group in the Department of Human Sciences at Loughborough University has been working in the general area of computer ergonomics for more than ten years. To many people the discipline of ergonomics in relation to computers is concerned with issues such as the visual characteristics of the display, along with the optimization of features of the hardware, the physical environment, and the workplace. However the members of HUSAT have long been aware that this 'traditional ergonomics' approach is completely inadequate when considering the complex socio-technical systems that have arisen through the introduction of computers into so many aspects of the working environment.

It cannot be emphasized too strongly that a consideration of the human factors of computer systems must include not only these traditional ergonomic issues but also the design of jobs, the ergonomics of dialogues, the provision of training and the planning and implementation of technological change. At all times the focus of attention of the computer ergonomist must be upon the people in the system and all the factors that affect them.

It is all too easy to fall into the trap of measuring that which is relatively easy to measure but ignoring that which is less easy to quantify. Thus many ergonomics guidelines contain prescriptive statements about the hardware and the physical environment that abound in numbers. These recommendations can, and in Britain, I believe, do, divert attention from the less quantifiable but no less important issues such as the software interface and the design of jobs.

I do not wish to suggest that we have all the answers to the problems related to vision, hardware, or the physical environment; indeed this book is testament to the fact that we do not. I do suggest, however, that there are many areas of computer ergonomics in which the necessary research has hardly begun.

Thus in attempting to place some perspective upon the issues of possible VDU health hazards the first point I would like to make is that the research that is required to investigate these possible health hazards is only a small part of the immense research effort that is needed in computer ergonomics.

It is my belief that it is not enough simply to undertake the research. We must ensure that the solutions that arise from the research into various computer ergonomics problems are implemented. Thus in HUSAT we not only undertake research but also consultancy and education in computer ergonomics. The penalty for this concern for the application of research is that immediately one leaves the research environment and enters the real world, one faces people asking awkward questions to which we do not, at present, have the answers. However, we do have some knowledge about many of the complaints and problems but all too often we see that this knowlege has not been utilized. I suggest that if a fraction of the effort that has been dissipated in the VDU health hazard scares of the past were to have been directed towards the dissemination and application of existing ergonomics knowledge the productivity and quality of working life of many VDU operators would by now be greatly improved.

From the comments I have made so far it might appear that I consider the possibility of VDU health hazards to be of little consequence. I assure you this is not so and to illustrate the point I would like to describe briefly how this book came about.

If a tradition can be established in three years then it might be said that it is traditional for HUSAT to hold a conference in December on some aspects of computer ergonomics. In December 1978 Tom Stewart, one of the original members of HUSAT, organized what was initially planned as a small scientific meeting on 'Eyestrain and Visual Display Units'. This escalated into a one-day conference attended by over 300 people at the University of Loughborough on 15 December 1978 (Stewart, 1978).

In December 1979 Ian Galer from the Department of Human Sciences at Loughborough University and I jointly organized a one-day meeting on the ergonomics of computer dialogues (Galer and Pearce, 1980). The 1980 December HUSAT conference was originally intended to review the issues and the progress that has been made in computer ergonomics over the last few years. However, events overtook these plans.

I will explain in some detail the origins of the contents for this book, as I believe it illustrates some of the problems associated with the investigation of such sensitive issues as possible VDU health hazards and also introduces some of the contributors.

In March 1980 I attended Professor Grandjean's international scientific workshop in Milan on the Ergonomic Aspects of Visual Display Units (Grandjean and Vigliani, 1980) to give a short paper on British Trade Union Guidelines relating to Visual Display Units. During this workshop Dr.Tjønn gave an unscheduled presentation in which he described the facial rashes that had been reported amongst a number of VDU operators in Norway. He asked the delegates to the workshop whether they had seen anything similar in their own countries. Dr Tjønn acknowledged that those involved in investigating the problem were, at that time, mystified as to the cause of the problem or its solution.

In short, Dr Tjønn had presented what appeared to be a *prima facie* case of a possible new VDU health hazard. I later became aware that these facial rashes had received significant publicity in the national press of various Nordic countries but to my knowledge nothing of a similar nature had ever been reported in the UK.

In Milan I was aware from the list of delegates that although there were a number of British users and computer manufacturers present there was no representation at the workshop from the British Health and Safety Executive nor, for that matter, from any of the British trades unions. I returned from the conference very concerned that what had been reported in Milan by Dr Tjønn might be misreported, directly or indirectly, in the UK.

This concern has its origins in the many conversations I have had with unions, VDU operators and managers of VDU installations in the UK. I have frequently been called in by both management and unions to investigate various ergonomic problems related to using VDUs. During these investigations I am always asked questions about the effects that working with a VDU may have upon the operator. A common feature of these questions is that some operators genuinely fear what working at a VDU will do to them.

Thus I and a number of other ergonomists also in Milan were worried that what was reported in Milan in March as ten possible cases of facial dermatitis amongst VDU operators might be misreported in the UK in April as, for instance, a hundred cases of skin cancer. Lest anyone choose to misinterpret this statement may I emphasize that we have no grounds whatsoever for suggesting any relationship between using a VDU and the development of a skin cancer.

Thus on my return to the UK I informed a representative of what I believed to be the relevant authority in the UK (Colin Mackay of EMAS) of the statements made by Dr Tjønn. I also wrote an article (Pearce, 1980) in one of the weekly computer newspapers and several academic journals reporting the conference and Dr Tjønn's appeal for reports of similar cases in other countries. I emphasized the limited evidence available at the time and the need for a thorough investigation of the facts. I was only too well aware that a sensational report could precipitate a new VDU health hazard scare. I might add at this point that my article was reproduced unedited by the computer newspaper. Unfortunately I had no influence over the headline or the cartoon that accompanied the article.

In the meantime I had been attempting to contact other individuals in the UK to whom I could refer any queries as I have no medical knowledge of dermatitis. From several sources I was referred to Dr Rycroft and his colleagues at St John's Hospital for Diseases of the Skin. Subsequent correspondence suggested that a small number of cases of facial skin problems amongst some male VDU operators had been investigated in the UK as long ago as 1978. While investigating who else might have relevant knowledge that could help to quantify the problem related to this possible VDU health hazard my attention was drawn to a paper given by Dr Zaret in Paris (Zaret, 1980). Dr Zaret acted as a consultant to the Newspaper Guild of New York in 1977 after two *New York Times* employees who worked at VDTs developed cataracts. In the paper Dr Zaret presented in Paris he suggested that in the light of new facts he had informed the Guild that he had changed his opinion, and now believed that there was a greater danger for VDT operators than there was three years ago, when he first investigated the problem. Thus towards the middle of 1980 a number of serious questions had been raised as to the possibility of direct health hazards to the operators of VDUs.

The number of people affected was apparently small, and there was no evidence of any previously unrecognized direct health hazard that could affect all VDU operators. It seemed likely that a number of individual operators were highly sensitive or predisposed towards developing these symptoms. However, I believed that to dismiss these few reports as being of little consequence would be irresponsible. These issues needed to be investigated.

This raises the question as to who or which organization should be responsible for such an investigation. Indeed, the question might be raised in a more general way—who is responsible for the avoidance of the problems associated with the introduction of VDUs in particular and of new technology in general?

I have no doubt that the manufacturers and purchasers of computer equipment must take a much greater responsibility for the application of ergonomic principles to their equipment and the workplaces. Manufacturers must put ergonomics into their equipment as well as into their brochures. Purchasers, with the aid of manufacturers, must apply ergonomics to the physical and psychological environment of the workplace.

However, I am not sure that we can leave the investigation of possible direct health hazards solely at the door of manufacturers and purchasers. However conscientious they might be, I do not believe that manufacturers and managements have the resources to conduct such investigations. Even if they do, will they be believed to have produced impartial results? I have already heard the *VDT Manual* (Cakir *et al.*, 1980) being criticized as a 'management document' because its production was originally funded by the European Newspaper Publishers Association.

Thus I believe that some independent or government body has to take the initiative. I am in no way questioning the integrity or technical capability of the individuals from the various occupational health agencies in the UK nor am I in a position to question the priorities given to VDU problems over other occupational health issues. However, I have already mentioned that a small number of facial skin problems amongst VDU operators were first identified in

Britain in 1978. Why has this information not been published until now? I believe that this begs the question as to whether the resources and organization of the various government agencies are adequate to undertake or co-ordinate the investigation of the human factors problems associated with the introduction of new technology. It is of little comfort to operators of VDUs to suggest that these possible direct health hazards of VDUs will only affect a small number of individuals who are highly sensitive or predisposed towards developing these symptoms if we are unable to define the nature of this sensitivity or predisposition.

An earlier VDU health hazard scare related to epilepsy was neatly placed in perspective by Dr A. Wilkins (1978) at the conference in Loughborough in December 1978. He not only indicated the proportion of epileptics who are photosensitive but also identified the epileptogenic attributes of VDUs. As yet I believe that we are unable to quantify the proportion of the population of VDU operators who might be affected by these two possible health hazards, if indeed they can be proved to be related to the use of a VDU.

Even if we could quantify the proportion of operators who might be affected, this does not remove the fears and worries of the operators. To quote relative occupational injury or occupational disease statistics may help to apportion resources in their investigation but it rarely reduces the concern amongst the section of the population that perceives itself to be at risk. Even if the actual risk is low, the perceived risk may be far higher, especially when dealing with highly emotive issues such as possible VDU health hazards that may affect an operator's sight, appearance, and well-being. One only has to look at the accidental death and injury statistics for an average day in the UK to realize that the amount of publicity and concern about the different types of accident is out of all proportion to the injuries or fatalities involved. Accident statistics are notoriously difficult to compare due to under-reporting and differences in definition. However, it is clear that more people are killed or seriously injured every day in the home than on the roads of the UK. Fatal or serious accidents occurring in the home and on the roads combined exceed the fatal and serious accidents occurring at work by a factor of ten.

I suggest that the publicity and concern regarding possible VDU health hazards is also out of proportion to the possible risk, but quoting statistics does not make the fears and worries just disappear. We must identify the causal factors.

One of the factors that influence the likelihood of an occupational disease or injury is the amount of time that the worker is exposed to the source of the risk. One of the arguments used by trade unions for the reorganization of work involving the use of VDU is to reduce the amount of time 'on screen' thereby reducing the exposure time of the employee to the VDU and the possibility of damage. In the absence of relevant research I believe that this position is entirely justified. The unions seek protection for their members in the UK by including statements in their policy documents and new technology agreements that specify maximum 'on-screen' time (usually at 4 hours) and rest-pause criteria. There are a number of other arguments used to justify the various rest pause criteria that have been negotiated. However, I believe that it is wholly inadequate to seek rest pauses or maximum on-screen time to compensate for fatigue or to minimize the risk of damage to the operator. The source of fatigue and the source of potential damage should be eliminated.

However, it is clear that while there exists the possibility of direct or indirect health hazards, unions will continue to seek agreements that reduce the possibility of fatigue or injury to their members.

If one accepts the argument that it is the perceived risk that influences the level of concern expressed and that this source of worry will only be dissipated by a thorough investigation, the problem arises as to what constitutes a possible VDU health hazard. When does the rumour of a health hazard merit detailed investigation? There is no simple answer to this question. The resources required to investigate every spurious rumour and allegation would be immense. It might be suggested that there has to be some face validity to the allegation for it to merit investigation, but even this is difficult to define. One could imagine a headline that announces 'VDU causes baldness' and cites as evidence the number of bald male VDU operators in a particular organization. Does this warrant serious investigation or simply a denial of any causal relationship? Will the denial be believed? It is much easier to suggest a causal relationship and to prove that it exists than it is to prove that a causal relationship does not exist. In many ways once the allegation has been made or the rumour rife the damage has been done.

The question arises as to whether an investigation of the rumour and the consequent publicity adds credence to the rumour and spreads it wider. An illustration of this problem is provided by the latest VDU health hazard scare that has come out of Canada. I quote from a Reuter news story. 'Ottawa, Aug. 6, Reuter—The Canadian Health Department has said it will not test any more video display terminals at any news organizations because it feels no hazards are involved. A controversy over video terminals, television screens linked to typewriter-like keyboards and widely used by newspapers and news agencies, developed after four women at the *Toronto Star* newspaper gave birth to children with abnormalities. All four had regularly worked on video display terminals at the *Star*. The Ontario Province Health Ministry tested the terminals and reported the birth abnormalities *did not seem* [my emphasis] to have been caused by radiation from the machines. Yesterday Walter Zuk of the Federal Radiation Protection Bureau said the Department would not divert staff from work on genuine radiation hazards to look at video machines which they know emitted no dangerous radiation.'

This denial followed a number of reports in the Canadian press alleging a relationship between the use of a VDU and the subsequent birth to a number of operators of children with abnormalities.

Many of the health scares surrounding VDUs have suggested the possibility of radiation hazards. The *Toronto Star* commissioned a report from the Radiation Protection Branch of the Ontario Ministry of Labour (Aiken, 1980). This included the monitoring for radiation emission of all 296 operational VDUs in the *Toronto Star*. No X-ray, radiofrequency, or microwave radiation was detected in excess of normal background levels. However, it has been suggested that some of the radiation emission surveys conducted in various countries have neglected to measure emissions in certain important parts of the spectrum. It has also been suggested that these surveys have only been conducted on new VDUs or those working properly and that surveys are needed of malfunctioning VDUs.

The Canadian Centre for Occupational Health and Safety also produced a review of the published literature on VDU health hazards (Purdham, 1980). The summary of the review states 'the published data would indicate that generally emission levels from VDTs are low, in some parts of the electromagnetic spectrum being orders of magnitude below current standards. However in parts, the work is poorly documented with respect to raw data and methodology'. The latest radiation emission survey in the UK is from the Health and Safety Executive and is presented by Mr Ernie Cox in Chapter 5.

There are immense problems of mounting a conference on such sensitive issues as possible VDU health hazards. I imagine there are some who view a conference on two particular, possible VDU health hazards as an unnecessary and misguided exercise that interferes with the technological progress and economic logic of computerization. Others might seize upon the issues and adopt a Luddite stand. I cannot support either of these approaches. I believe that it is vitally important that where there is *prima facie* evidence of a health hazard the issues must be considered in a responsible and rational manner.

To develop a constructive approach to these complex problems requires the co-ordination of resources such as those of the Health and Safety Executive, the National Radiological Protection Board, and the Employment Medical Advisory Service. However, the co-ordination of these resources will not suffice to achieve effective utilization of new technology. There is an important requirement to supplement the activities of these bodies since their primary purpose is to avoid injurious or deleterious effects of technology. This responsibility, significant as it is, does nothing to ensure that human well-being and performance are positively improved through appropriate application of new technology. The vital need for the next decade is for successful application of human factors and ergonomic principles integrated with relevant medical and associated expertise. There is a need for one agency solely concerned with the human aspects of new technology that can co-ordinate these resources; one agency that not only promotes ergonomics in industry to avoid the indirect health hazards but can also respond quickly and authoritatively to allegations of direct health hazards; a body that encourages the known solutions to be applied and can co-ordinate both nationally and internationally the research effort that is desperately required.

If such a body were to be established it is vital that it embraces or could call upon the whole range of expertise required. It must not concern itself solely with the possible direct health issues of new technology but must also consider the indirect health issues and the quality of working life of those individuals who are to work with new technology. New technology should be our servant. Far from being a source of potential injury or damage it should enhance our environment in the broadest sense.

REFERENCES

Aiken, H. (1980). 'Radiation tests on VDTs at the "Toronto Star" ', Memorandum dated 8 August to Assistant Deputy Minister, Occupational Health—Safety Division, Ontario Ministry of Labour.

Cakir, A., Hart, D.J., and Stewart, T.F.M. (1980). *Visual Display Terminals*, John Wiley, Chichester.

Galer, I., and Pearce B.G. (eds) (1980). 'Man–computer communication', *Ergonomics* **23**, 9.

Grandjean, E., and Vigliani, E. (eds) (1980). *Ergonomic Aspects of Visual Display Terminals*, Taylor & Francis, London.

Pearce, B.G. (1980). 'Staring in the face of a big small screen scare', *Computing* **8**, No. 16, 17 April.

Purdham, J. (1980). *A Review of the Literature on Health Hazards of Video Display Terminals*, Canadian Centre for Occupational Health and Safety, Hamilton, Ontario.

Stewart, T.F.M. (ed.) (1978). 'An edited transcript of the one-day meeting on Eyestrain and VDUs', The Ergonomics Society, the Department of Human Sciences, Loughborough University, Loughborough, UK.

Wilkins, A. (1978). 'Epileptogenic attributes of TV and VDUs', in Stewart, (1978). An edited transcript of the one day meeting on Eyestrain and VDUs, The Egonomics Society, Department of Human Sciences, Loughborough University, Loughborough, UK.

Zaret, M. M. (1980). 'Cataracts following use of cathode ray tube displays', Paper presented to the International Symposium of Electromagnetic Waves and Biology, France, 30 June–4 July 1980.

Chapter 3

Facial Rashes Among Visual Display Unit (VDU) Operators

R. J. G. RYCROFT AND C. D. CALNAN

St John's Hospital for Diseases of the Skin, London, UK

Visual display units (VDUs) had been in use for short periods in an office at a factory for three years. However, in August 1978 two operators began to work full time, every day, on two VDUs. The office was a large open-plan building with approximately sixty people employed in it. It was fully air-conditioned and had fluorescent lighting.

THE PROBLEMS

After a month or two, one of the two full-time operators complained of red patches on the malar areas of his cheeks, the areas over the zygomatic arches. Approximately two weeks later the other full-time operator complained of a similar skin problem. Two other men subsequently became affected, one working full-time on VDUs and the other working only part-time.

Several other offices at the same factory also had the same make of VDU. Many other employees therefore used VDUs but almost all for short periods only. In one office there was a woman who worked full-time on the same make of VDU and had no symptoms at all. None of the men had any lesions to show at the time the factory was visited, but these were described by the medical officer as consisting of erythema and papules on the face, somewhat analagous to rosacea.

ATTEMPTED SOLUTIONS

It must be mentioned that the operators in this office appeared to be thoroughly sensible, well-motivated, hardworking employees, whose main aim was to get on and do their job. There did not appear to be any aggressive feelings towards the introduction of the new system; everyone seemed to be in favour of it. The employees themselves had made some attempts to solve the problem. The position of the VDUs was changed so that the operators were dependent more upon daylight than upon fluorescent lighting, but this made no difference.

The first operator moved his VDU so that it pointed directly at a wall and he viewed it obliquely from the side. When he did this, he found that he had no skin trouble and he continued to work satisfactorily for several months. This operator had also found that using one of his wife's moisturizing creams on his face two or three times a day helped to protect him.

The second operator put a piece of ordinary window glass in a frame in front of his VDU and found that this was entirely satisfactory. Whenever he removed the glass, the skin trouble would recur within a few days.

The difficulty in studying this situation is that many changes tend to be made at once, which creates problems in assessing the effect of one particular change. This is not peculiar, of course, to the VDU environment; it happens with any health hazard outbreak at a place of work. Many changes tend to take place within a relatively short period.

INVESTIGATIONS

Consideration was given to several factors as possible explanations of these rashes, including static electricty and low humidity. There was a substantial amount of static electricity from nylon carpets in the department. The carpets were, however, treated specially for this every six weeks. The relative humidity in the air-conditioned office had been around 40%, but this was said to have been raised to between 50% and 60% a month previously. It was uncertain whether this increase had had any beneficial effect.

The National Radiological Protection Board carried out an ionizing and non-ionizing radiation survey on one of the VDUs involved. Using an International Light IL 730 Actinic Radiometer, providing an integrated reading over the wavelength range 200–315 nm, weighted for special effectiveness in accordance with the ACGIH TLV for ultra-violet (UV) radiation, it was found that when the screen was full of characters the measured amount of UV at the screen surface was less than 10^{-7} W cm^{-2}. The UVA radiation (315–440 nm) was found not to be above the 'background', which is substantially less than 1 mW cm^{-2}.

DISCUSSION

No link has been proved between the VDUs and the facial rashes reported. The evidence for connecting the facial rashes with the VDUs is entirely circumstantial. It is because this circumstantial evidence appears to be so strong and because no alternative explanation for these rashes has yet been forthcoming that this report is made.

ACKNOWLEDGEMENTS

I must acknowledge that my colleagues Dr Charles Calnan and Dr John Hawk were the doctors who actually investigated this particular outbreak described. I was not directly involved but have had discussions with them since they visited the factory and I have a fair grasp of what they saw and found.

Chapter 4

Report of Facial Rashes Among VDU Operators in Norway

HANS H. TJØNN, MD

Directorate of Labour Inspection, Oslo, Norway

During the winter of 1979 there were rumours in Norway that VDU operators became 'sunburned' during work. However, the medical profession as well as radiation physicists and terminal producers were sure that no significant radiations were being generated through the screens.

However, in March 1979 two operators working in the same room developed facial rashes. The factory doctor saw the two operators and, as he knew nothing about VDUs and skin reactions, he doubted that the operators' moderate reactions had anything to do with the working environment.

The two operators, both female, did not develop the erythema every day, but they both felt very strongly that the facial erythema was due to their work. They therefore complained to the safety delegate, who notified the organisation's work environment committee. From then on many steps were taken in order to investigate the circumstances surrounding these two cases.

First, measurements were made to see if the lighting tubes in the room were emitting ultra-violet light in such an amount that the operators in the room could possibly acquire a real UV sunburn. After investigation and measurements made by technical personnel this possibility was excluded. At this stage the question of the effects of video terminals on health were placed before the Labour Inspection in Bergen, and later communicated to the Directorate of Labour Inspection in Oslo.

We took action on different levels in order to explore and elucidate the circumstances surrounding the two cases.

(1) Dermatological expertise was obtained from the Department of Dermatology at the University of Bergen.

17

(2) Radiation expertise was obtained from the National Institute of Radi-
     ation Hygiene in Oslo.
(3) Factory doctors were asked to report cases of facial rashes which
     developed during working hours among VDU operators in Norway.
(4) Provocation tests were carried out in order to see whether or not the
     rashes could be reproduced under controlled conditions.
(5) Experiments were carried out in the room where the cases came to light in
     order to see what measures might prevent the skin reaction.
(6) The Christian Michelsen Institute in Bergen was called upon to see if
     static electric fields from the units, in combination with charged dust
     particles in the working atmosphere, could possibly be the cause.

The present status of the problem of facial rashes among VDU operators in
Norway is as follows:

(1) Doctors in Norway have, during 1979 and 1980, notified the Directorate
     of Labour Inspection of about forty cases of facial rashes among VDU
     operators. These cases have been routinely evaluated by competent
     dermatologists. After dermatological evaluation, sixteen cases are
     assumed to be facial dermatitis caused by factors related to working
     conditions.
(2) The predilection site of the dermatitis was the arcus zygomaticus region of
     the cheek. To a minor extent, but not in all cases, exanthema also
     developed at the point of the chin and at the forehead. Some cases also
     developed a more general redness of the face and also the uncovered skin
     on the front of the neck.

DESCRIPTION

The exanthema developed after less than 2 hours of work—up to after 5–6 hours.
The patients often described that before the outbreak they had a feeling of itching
or of being cautiously patted with a feather. Then the redness developed
successively without being very intense or especially embarrassing. Most cases
also developed minor desquamation and showed some small papules. The
exanthema most often disappeared during the first few hours after leaving work.
Most of the cases could demonstrate a normal skin the next morning. One or two
of the cases, however, needed more than eighteen hours to recover.

COMMON FACTORS IN THE WORKING ENVIRONMENTS

All reported cases had their outbreak during the winter months. The working
temperatures in the rooms were around 20–22°C. The relative humidity was
around or well below 40%. All working rooms where the skin trouble occurred
were covered with carpets of synthetic fibres except one room, where the floor

had a vinyl covering. There was a substantial amount of static electricity in the rooms where the operators developed their dermatitis and steps had been taken to prevent the build-up of static electric charges.

Investigations

Clinical investigation of the cases did not reveal any unsuspected or abnormal findings. Some of the cases could report atopic dermatitis in childhood or in relatives. Some of them were reported as having seborrhoeic skin in the face. None of the cases were reported as having atopic manifestations from the respiratory organ, eyes, or from the skin at the examinations. None of the cases which were grouped as having VDU dermatitis had symptoms indicating photosensitive dermatitis, except possibly in one case. Vitamin A in blood was determined in twelve cases. The results were interpreted as normal findings.

Provocation tests were carried out on some of the cases. One series of well-controlled provocation tests conducted by Dr Arvid Nilsen at the University Clinic in Bergen and covering five cases were deemed negative. Each case sat for six hours with one of the cheeks directed towards the screen. The five cases were each tested for three days. On the first day the unit was not connected to the mains electricity; on the second day the unit was connected, but was not otherwise in operation; on the third day the unit was filled with characters but not operated by the test subject. None of these cases developed exanthema.

Other provocation tests have, however, been carried out in other ways, i.e. cases which have showed skin reaction during working hours have been followed operating the unit under completely normal conditions, and the outbreak of redness, desquamation, and of papules are described.

Experiments have been carried out in order to prevent skin reaction among known cases.

(1) Placing a window-glass shielding between the operator and the screen did not prevent facial rashes developing. On the contrary, Dr Victor Linden reported that from his experience cases developed facial rashes very distinctly behind a glass shielding. He would not, however, confirm that the time of exposure was shorter or if the reaction was more distinct under this experiment. This point contrasts with the information given in the previous chapter.

(2) Placing of a glass basin filled with an electrolyte and connected to the surface of the unit and then grounded seemed, however, to prevent cases from developing an outbreak. Unfortunately, however, it was not possible to construct a basin without leakage. Several units were damaged because of this, although expertise was used to construct these aquaria.

(3) Replacement of floor carpets with anti-static carpets (metal-thread woven in the carpet and grounded) were carried out and this seemed to prevent cases from occurring.

(4) One case who was diagnosed as developing a facial rash when operating video display units was placed in front of a Van de Graaff machine by Dr Linden of Bergen. He reported that after some minutes the subject complained of itching and said she had the feeling of being touched by a feather. Some time later she developed redness over the zygomatic region and the experiment was discontinued. This experiment will, however, be repeated.

(5) In one office where three persons were working and two of them operating VDUs both operators developed facial dermatitis, and it was decided to replace the floor carpet with an anti-static one. The walls of the room were covered with vinyl wallpaper. After replacing the carpet the skin trouble became worse, and the two operators complained of increased discomfort and skin trouble. The industrial hygienist, who had conducted the environmental survey before the improvements were made, took measurements of the electrical conductivity of the new carpet. To his surprise he found that it was near zero. It was found that the new carpet was fastened to the floor with a substantial amount of isolating glue and grounding had not been made. After replacing the carpet and the isolating glue the proper grounding was secured and the skin manifestations and complaints disappeared. (Reported by Victor Linden, Sturla Rolfsen and Olav Faerovik.)

WORKING HYPOTHESIS

We believe in Norway that some very few VDU operators may acquire contact dermatitis. We are carrying out research in order to clarify the possible roles of static electricity, static electric field, and airborne dust particles in the generation of the observed skin reactions. This research is being conducted by research physicist Walter Cato Olsen at the Christian Michelsens Institute in Bergen, who is due to publish his first report in 1981. I will now describe some of the cases.

Case 1

A girl, aged twenty, developed psoriasis on her knees and elbows at the age of six or seven. She started VDU work in 1978 and from late autumn in 1979 she noticed itching and redness on her face after a full day's work. The dermatitis was localized over the cheekbones and the glabella and also at the tip of the chin. The dermatitis was described by the dermatologist as moderate. The girl observed that the dermatitis disappeared overnight and she had no symptoms during weekends or holidays. She did not feel the necessity of seeing her own physician about this skin reaction. The case came to light when the company doctor went to the workplace to inspect the VDU operators' faces at the end of a working day. Provocation tests on these operators in the same room turned out positive. As well as not seeing her own doctor about the rash she had also seen the company physician twice during the time that she had had the rash but she had not notified him of the rash.

Case 2

The second case had had no previous dermatological problems. She started VDU operation early in 1977 and when working she sometimes developed a feeling of itching on the face. She developed a slight redness and some telangiectasia. The dermatologist also identified some minute papules above the angulus mandibula at the border of the redness. The itching and the skin irritation did not bother this patient very much and I have no information about provocation tests in this case. The clinical examination did not reveal any special symptoms or signs of other diseases.

Case 3

This VDU operator is thirty-eight years old and she also had had no skin problems before working with visual displays from 1976. During the last two years at certain times she had been bothered with a redness in the face like sunburn. She felt that the redness was more intense when she felt the work was stressful. The symptoms were visible as early as 1–2 hours after starting work, but it did not develop every day. The symptoms might occur for two successive days or for as long as one week or more, but she had symptom free periods of various lengths. Provocation tests were not carried out in this case. The predilection site was over the cheekbones and the chin, and she also had some redness on her nose-tip. When not working, in the vacations and in summer, her skin was completely normal.

Case 4

A female aged thirty-two who had had no previous skin diseases started VDU operations in 1978 and developed dermatitis in November 1979. During the summer of 1980 she experienced no symptoms despite working in the same place. From 1 to 3 hours after starting work in the morning she felt an itching of the face and the exanthema developed over the cheekbones, around the eyebrows and on the chin. There was a little erythema over the eyebrows, on the upper lids, and on the tip of the nose.

Case 5

A VDU operator aged fifty-six, who had had some acne vulgaris started VDU operation in 1978 and after one week's work she had a feeling of glowing cheeks and said that her face felt uncomfortable. The dermatologist described a papulo pustulos exanthema with unsharp borders to normal skin over the zygomatic regions at the stern and the chin. She stopped working with VDUs, she would not go for a provocation test and she ceased to work for this company.

POSTSCRIPT

Oslo 26 October 1981

Through the winter of 1981 a few new cases of dermatitis associated with VDU work were reported in Norway. All cases were similar to those previously reported. The research work conducted by the Christian Michelsens Institute by Mr Walter Cato Olsen was documented in a report from the Institute in April 1981 (Ref. CMI No. 803604-1), and the results are summarized as follows:

(1) Unexplainable facial rashes are experienced by some operators of visual display units (VDUs).

(2) The results of a survey of electrostatic phenomena in a series of VDU work areas indicate that rash-prone operators are commonly exposed to extreme electrostatic fields. These are caused partly by high voltages associated with the display screens and partly by electric charges accumulated by the bodies of the operators.

(3) Direct measurements of aerosol deposition onto substrates attached either to the cheek of an operator or mounted on a specially designed high voltage sampling device show that the precipitation of submicron particles increases substantially in the presence of an electric field, and that the increase is nearly proportional to the intensity of this field. Under conditions equivalent to those reported to exist when rashes occur, the measured particle flux reaches values above $104 \ mm^{-2}hr^{-1}$, which is at least an order of magnitude higher than the flux at zero field conditions.

(4) It is suggested that field-enhanced exposure to submicron, irritant fractions of the ambient aerosol causes the observed rashes, and possibly also other health problems such as eye discomforts.

(5) The elements sulphur and chlorine, which are common constituents of the general, outdoor aerosol in the geographic region of the study, are relatively abundant in the field-generated deposits and may be components of the chemical compounds that are active in the development of the rashes.

After studying the report we believe that the facial rash which has been seen in Norway is in fact an occupational contact dermatitis, caused by irritating submicron dust particles precipitating on the skin of the VDU operator if he is accumulating static electricity.

Further research should be carried out in order to investigate whether or not red eyes in VDU operation might also be caused by submicron particle precipitations.

The recommendations given by the Directorate of Labour Inspection to prevent these effects in VDU operation seems to be adequate. These recommendations are the elimination of build-up of static electricity and the grounding of display terminals which might give rise to high electrostatic fields. The manufacturers of VDUs should also design the unit in such a way that the outside screen potential is nearly zero volts. Furthermore, the employer should be responsible for designing the VDU workplace so that the operator is not subject to a build-up of high levels of static electric charge on the body.

Chapter 5

Radiation Emissions from Visual Display Units

E. A. Cox

HM Principal Inspector of Health and Safety
Health & Safety Executive
25, Chapel St
London

INTRODUCTION

Widespread concern has been expressed over the possible exposure of operators to radiations emitted from visual display units (VDUs) based on cathode ray tubes. Such devices are increasingly being introduced into business premises to communicate with computer and processor systems. In this chapter the potential hazard of any electromagnetic radiation emitted by a VDU is considered. The Health and Safety Executive (HSE) will, I believe, be issuing a separate Technical Guidance Note to cover other aspects of VDUs such as ergonomics, design, siting, lighting, etc.

The question of potential radiation emissions from visual display units goes back several years and there have been a number of surveys conducted in various countries. However, of those available to us it appeared that there were areas in which there were doubts about the completeness of the data or the manner in which it was collected. Consequently to quantify any potential radiation emissions from VDUs the HSE made arrangements to carry out a comprehensive survey, following agreed survey protocol. In this survey measurements were made of the radiation in various regions of the electromagnetic spectrum from all known types of VDUs currently manufactured or marketed in the UK. It was decided that it would be better to try to cover the whole field rather than a 10% or 20% sample, because one set differs from another. Since the survey was clearly going to take some time it was felt that it would be better to take the measurements at the manufacturer's or supplier's premises using demonstration models, for no other sinister reason than we were then able to have as much time on those models as we needed for the survey. In most other work situations VDUs are kept busy and there may have been problems if we wished to spend a morning or more making the measurements. Also it is often difficult to discover where a manufacturer's products are in use. For these reasons it was decided to conduct the survey at the manufacturer's or supplier's premises.

Instead of a fairly short list of establishments as I had envisaged, the task became more and more formidable. In the end, with some selection, we drew up a list of sixty firms and measurements were made on more than two hundred different types of VDUs.

The next problem in the survey was to decide which parts of the spectrum to cover. We felt that the survey needed to be undertaken or at least got under way promptly, and for that reason we used the National Radiological Protection Board under an agency agreement to carry out the field measurements for us. An agreed survey protocol was established for the measurements concerning the location where the measurements were to be made and the instruments to be used. The choice of instruments was limited to those already available to the NRPB and from their list we selected instruments which would cover a large sector of the electromagnetic spectrum. This selection omitted one or two areas which were not covered by that instrumentation and we attempted to fill those gaps with other experiments mentioned later in the chapter.

The protocol for the survey is set out in Appendix A and describes the instruments used and the positions. The measurements were made in contact with the screen or the casing on the five accessible sides. Other surveys have made estimates of the worker's position relative to the VDU, but this is open to some contention. These readings will vary according to whether one is short-sighted or whether sitting close to another working VDU. It was decided to make the measurements at contact or as near as contact as possible with the surfaces. Clearly, this maximizes the readings. To maximize them further the controls of the VDU were adjusted to their maximum setting consistent with an adequate display, the screen being filled with 'M' characters.

After data collection we had to decide what to do with the data and how to disseminate the information. Reports including all measurement data on each VDU surveyed have been supplied to all the manufacturers/suppliers in respect of their equipment. Because of the constraints of Section 28 of the Health and Safety at Work Act 1974 specific results cannot be quoted here, although individual firms can, if they wish, release any information about their products. I have selected from the data the numbers that are high within the survey, but I cannot discuss any particular firm's products.

I will summarize and discuss the implications of the results of both this main field survey and of some other special measurements made on a smaller sample of selected VDUs. The results are compared, as far as we were able, with the permissible standards available for continuous occupational exposure. That is easier to say than to do, of course; there are standards in a number of areas of the electromagnetic spectrum but other areas have virtually no standards. We had to resort to draft standards and suggestions because there are no other standards with which to compare the results.

SUMMARY OF RESULTS

Tables 1 and 2 summarize the results given in this section.

X-ray Radiation

The maximum X-ray radiation from any surface of the VDUs surveyed was less than 10 μrem hr^{-1}. This is fifty times less than the permitted maximum exposure rate for household electronic apparatus of 0.5 mrem hr^{-1} at 5 cm laid down in BS 415:1979. (The standards referred to are given in Appendix B.) The emission may also be compared with 750 μrem hr^{-1}, which is the radiation dose rate level at which persons are required to be designated as 'classified' under the Factories Act Radiation Regulations. These measurements were made in contact with the screen and casing sides. Two instruments were used, one to detect 'hot spots', i.e. small areas of enhanced dose rate, and the Victoreen 440 RF/C, an instrument designed to look at devices with a potential for radio frequency interference. This is a notorious problem with many Geiger tube devices. The measurements showed nothing in terms of radiation dose rates and were similar to levels of natural background.

Ultra-violet Radiation

Measurements of ultra-violet emissions were made using a scanning spectral radiometer for wavelengths in the range 240–400 nm. No ultra-violet radiation was detected from any VDU at a wavelength less than 336 nm. The maximum emission from any VDU was 124 mW m^{-2}. Eighteen VDUs had an emission between 10 and 100 mW m^{-2} and all the rest gave less than 10 mW m^{-2}. 100 mW m^{-2} corresponds to 1% of the threshold limit value (TLV) for continuous occupational exposure in this region of the spectrum recommended by the American Conference of Governmental and Industrial Hygienists (ACGIH), a standard endorsed by the Medical Research Council (MRC). We and the NRPB work to this standard, which is also accepted within the WHO. The emission from the one set with the highest reading of 124 mW m^{-2} is well under the 10 mW m^{-2} laid down in that standard. The measurements were made with a Scanning Spectral Radiometer which scans across the spectrum picking up a line by line 4 nm bandwidth and integrates up the total power found in that

Table 1. Summary of radiation measurements on visual display units

Electromagnetic spectrum region	Maximum measured emission	Measuring instrument	MPEs, MELs, TLVs etc	Standard	Remarks
X-rays $\lambda < 100$ pm	dose equivalent rate <10 μrem hr^{-1}	Victoreen 440 RF/C	0.5 mrem hr^{-1}	BS 415:1979	Photon energy > 10 keV
Ultra-violet 240–336 nm 336–440 nm	irradiance 0 124 mW m^{-2}	NRPB Scanning Spectral radiometer (SSR)	— 10 W m^{-2}	ACGIH	Endorsed by MRC (private communication)
Visible 400–700 nm	irradiance calc radiance 2.5 W m^{-2} 0.8 W m^{-2} sr^{-1}	EG & G type 550 radiometer or UDT 40X opto-meter (with filters)	$21C_3$ W m^{-2}sr^{-1}	Draft IEC/TC76 Draft BS 4803	MPEs for viewing laser by diffuse reflection $C_3 = 1$ far$\lambda = 400$–500 nm $\log C_3 = 0.015$ (λ –550) far $\lambda = 550$–700 nm
Infra-red (near) 700–1050 nm	irradiance 50 mW m^{-2}		100 Wm^{-2}	ACGIH	Intended TLV
Infra-red (far) 10 μm–3 mm	4 W m^{-2}	Plessey pyroelectric detector	—		Measured irradiance equivalent to thermal radiation from body at 32°C. No measurable nonthermal radiation
Microwave and radio frequency 18 GHz–300 MHz	5 W m^{-2}	Narda 8300	100 W m^{-2}	ACGIH	See Table 2 for detailed measurements in region.
3 GHz–10 MHz	<1000 W m^{-2}	Raham 12	10 W m^{-2}	79-EHD-30	Spurious measurement. Not confirmed by ERA scan (Table 2)
220 MHz–10 kHz	electric field intensity >300 V m^{-1}	IFI EFSI	60 V m^{-1} 1500 V m^{-1} 3000 V m^{-1}	79-EHD-30 Draft foreign Draft foreign	220 MHz–10 MHz 30 MHz– 3 MHz 3 MHz–10 kHz

Table 2. Summary of radiation measurements on visual display units by ERA (microwave and radio frequency)

Electromagnetic spectrum region	Maximum measured emission Electric field intensity (V m^{-1})	Measuring Instrument	MPEs, MELs, TLVs etc (V m^{-1})	Standard	Remarks
Microwave and radio frequency					
300 GHz–30 GHz	Not measured		200	ACGIH	Endorsed by MRC 70/1314 (200 V m^{-1} = 100 W m^{-2} in free space)
30 GHz–1 GHz	Not measured				
1 GHz–300 MHz	< 0.01	Schwartzbeck VUME 1520A	200	Draft foreign	To be published for comment
300 MHz–30 MHz	0.18				
30 MHz–10 MHz	0.12	Schwartzbeck FSME 1515	60	79-EHD-30	Canadian standard for general public
10 MHz–3 MHz	0.5		1500	Draft foreign	
3 MHz–300 kHz	16				
300 kHz–150 kHz	600	Stoddart NM12AT			
150 kHz–10 kHz	1800		3000	Draft foreign	

region. Perhaps it should be noted that Mr Allen (NRPB), who has done a number of these surveys, has pointed out that significantly higher readings could be obtained from normal fluorescent lighting than from any of the VDUs examined.

Visible and Infra-red Radiation

There are various standards set for visible radiation. We chose to use the one contained in the Draft Laser Standard, which is fairly well researched and probably the most up to date, as it is about to be published. Indeed it seemed more restrictive than one or two other Standards available. Comparing the numbers with the standard the VDUs are still within the permissible level. I would stress that these measurements were taken in contact with the screen and therefore represent a much higher value than would be collected by an operator at any normal position. In the visible light region (400–700 nm wavelength) the maximum emission measured was 2.5 $W m^{-2}$. The maximum permissible exposure (MPE) for continuously viewing the diffuse reflection of a laser beam in the wavelength range 400–700 nm is given in the draft IEC and draft BS 4083 as $21 \times C_3$ $W m^{-2} sr^{-1}$, where C_3 is 1 for wavelengths (λ nm) from 400 to 550 nm and 10 to the power 0.015 (λ –550) for λ from 550 to 700 nm. This MPE is at least twenty-five times the maximum value of 0.8 $W m^{-2} sr^{-1}$ measured.

In the near infra-red region (700–1050 nm wavelength) the maximum emission measured was 50 mW m^{-2}, which is more than two orders of magnitude less than the ACGIH intended TLV of 100 W m^{-2}.

In the far infra-red region (10 μm to 3 mm wavelength) in a separate series of laboratory tests on a number of VDUs no radiation was detected apart from that due to a hot body. Since the working temperature of the VDUs tested never exceeded 32°C the measured irradiance of 4 W m^{-2} was less than that emitted by a human hand.

Microwave and Radio Frequency

Field measurements in the microwave and radio frequency regions of the electromagnetic spectrum were made with three instruments whose individual broad band responses overlapped, so that readings could be taken in the frequency range 18 GHz to 10 kHz.

In the survey one VDU gave an irradiance response of 5 W m^{-2} on the Narda 8300 instrument; all the rest gave less than 1 W m^{-2}. Over the frequency range covered by the Narda (18 GHz to 300 MHz) the ACGIH recommended TLV is 100 W m^{-2}, which is twenty times the maximum measured.

Thirty-six of the VDUs gave responses in excess of 1 W m^{-2} on the Raham 12 instrument, fourteen gave responses greater than 100 W m^{-2}, and the maximum approached 1000 W m^{-2}. However, the more detailed measurements described later did not confirm these results. The high value could have been obtained because of instrument sensitivity to radiation below the designed frequency range of 3 GHz to 10 MHz. Over the frequency range 3 GHz to 10 MHz the Canadian standard 79-EHD-30 gives a maximum exposure level (MEL) of 10 W m^{-2} for the general public, but at lower frequencies other standards permit much higher exposure levels.

One hundred and four of the VDUs gave electic field intensities in excess of 10 V m^{-1} on the EFS 1 instrument; thirty-nine of these gave intensities of over 200 V m^{-1} and the maximum found was over 300 V m^{-1}. No one standard spans the whole of the frequency range covered by the EFS 1 (220 MHz to 10 kHz). The Canadian standard 79-EHD-30 gives a MEL of 60 V m^{-1} for the range 220 MHz to 10 MHz. A draft foreign national standard to be issued shortly for public comments gives limit values of 1500 V m^{-1} for the range 30 MHz to 3 MHz and 3000 V m^{-1} for the range 3 MHz to 10 kHz. Again, detailed measurements show that the high measured value occurs in a lower frequency range than that to which the MEL of 60 V m^{-1} applies.

Microwave and radio frequency exposure standards are dependent on the specific frequency but the three instruments used in the survey possessed broad band responses. Because of this, supplementary measurements were made on four of the many VDUs that had given electric field intensities in excess of 300 V m^{-1} in the frequency range 200 MHz to 10 kHz. These measurements, which included a full frequency spectrum scan, were carried out by Electrical Research Association (ERA) Technology Ltd.

The four VDUs on which measurements of electric and magnetic field intensities were made at ERA gave very similar results. Since the magnetic and electric fields were found to be equivalent, only the maximum electric field intensities measured in each frequency range will be quoted.

The fields were measured on the surface of the VDUs close to the line output transformer. High field intensities in excess of 1000 V m^{-1} were measured at the fundamental and harmonics of the line scan frequency up to about 150 kHz. As the frequency increased from 150 kHz to about 3 MHz the maximum intensity fell rapidly to under 1 V m^{-1}. Above 3 MHz most of the field was caused by logic circuit noise and its intensity fell from 0.5 V m^{-1} to less than 0.2 V m^{-1} at 30 MHz. The measurements are given in Table 2. These specific results are consistent with the results obtained in the survey using the EFS 1 instrument but confirm that those obtained using the Raham 12 were spurious.

As the four VDUs investigated at ERA gave very similar results and were those that had given the highest intensities in the survey then the maximum measured intensity is likely to be greater than that from any other VDU microwave and radio frequency radiation emission. Exposure standards are currently under review in a number of countries but no one standard covers the whole frequency range. The standards compared with the measure intensities are given in Table 2; from this table the measured intensity in the frequency range 150 kHz to 10 kHz is 60% of the limit value.

In most other frequency ranges the measured intensities are one or more orders of magnitude less than the appropriate standard limit values. This is one area where there is particular difficulty with the standards. We have quoted from a draft Continental standard which I have received in confidence which is why I have not described its origin. The ACGIH are considering bringing out a standard for this part of the spectrum. Again these measurements were made in contact with the set. The signal fell off rapidly at distance, and at operator position was virtually undetectable; in the order of 1 V m^{-1} about 18 inches away from the set. Once again I do not wish to rely on distance measurements but we tried to quantify how the signal fell away and those are roughly the results.

CONCLUSIONS

The measured radiation emissions from correctly operating VDUs are much less than the limits for continuous occupational exposure in the various regions of the electromagnetic spectrum given in many national and international standards.

We have looked at a variety of VDUs and have operated them under extreme conditions. We tested demonstration sets, some would have been new, others would have been in use for considerable periods of time. The measurements were made in contact with the screen and the case and in those positions the permissible levels are not exceeded or even approached. Therefore it is very difficult to see that any significant radiation emission is coming from these devices.

The measured values are in excess of those to which an operator would be exposed since the measurements were all made with the brilliance and contrast controls adjusted to the maximum setting consistent with an acceptable display and very close to the surface of the VDU. In particular the signals in the radio frequency region of the spectrum were highly localized and reduced by a factor of at least a thousand at the operator's normal working position. Hence, the conclusion must be that the radiation normally emitted from a VDU does not pose a hazard to operators either in the long or short terms.

APPENDIX A: Protocol for surveys of Visual Display Units (VDUs) made on behalf of the Health and Safety Executive

1. Information to be asked of the Manufacturer/Agent

1.1 Make of VDU
1.2 Model identification
1.3 Serial number
1.4 Tube type and manufacturer
Tube ratings (EHT, beam current)
Screen phosphor type
1.5 Details of models not available at time of visit

2. Instruments to be used

2.1 **X-ray radiation measurement:** for photons of energy >10 keV
Victoreen 440 RF/C
Mini-monitor type 5.40 with type 5.42 X-ray probe
2.2 **Ultra-violet radiation measurement:** for photons of wavelength 240–400 nm
NRPB Scanning Spectral Radiometer (SSR)
2.3 **Ultra-violet A, visible and infra-red radiation measurement:** for photons of wavelengths 315 to 1050 nm
EG & G type 550 radiometer, with 315, 400, 550 and 700 nm filters
2.4 **Microwave/radio frequency radiation measurement:** for the frequency range 18 GHz to 10 kHz
Narda 8300 broad-band isotropic monitor (300 MHz to 18 GHz)
Raham 12 (with 82 probe) broadband monitor (10 MHz to 3 GHz)
Instruments for Industry (IFI)
EFS 1 electric field strength monitor (10 KHz to 220 MHz) with auxiliary handle

3. Method of Measurement

Check background reading on all instruments before switching on VDU equipment. Switch on VDU, fill the screen with a display using the letter 'M', and adjust user controls, i.e. brightness and contrast, to maximum consistent with an acceptable display.

3.1 **X-ray radiation**
3.1.1 Use the 5.42 probe to search around, and in contact with, all accessible surfaces. Note the nature of any support material between probe and instrument base. Check with and without a 1 mm thick lead sheet over the end of the probe to indicate any spurious RF interference. Record

highest genuine reading obtained in each compartment of a 3×3 grid, for each surface, as described in Section 3.4 below.

3.1.2   Make one measurement at all accessible surfaces with the 440 RF/C in contact with the surface at the area of greatest 5.42 probe reading for each surface

3.1.3   Make a measurement at 30 cm from the centre of the screen using the 5.42 probe (with and without lead sheet)

3.1.4   Mini-monitor measurements are to be converted to mrem hr^{-1} assuming that the effective energy of the radiation emitted corresponds to the EHT applied to the tube

3.2   **Ultra-violet, visible and infra-red radiations**
Where reasonable, minimize ambient levels by reducing local lighting and fitting black cloth or paper as local screening between the VDU and the SSR or EG & G 550.

3.2.1   SSR
Make one spectral scan at 4 nm intervals from 400 nm to 240 nm with the SSR orifice directed at the centre of the screen and in contact or as close as practicable. If not in contact, record distance

3.2.2   EG & G 550
Use the EG & GG 550 radiometer as a search instrument in contact with the screen and record the highest reading obtained in each compartment of a 3×3 grid as described in sub-section 3.4. Determine the position of maximum reading with the filter holder attached to the instrument and repeat the measurement with 315, 400, 550, and 700 nm filters used successively

3.3   **Microwave/radio frequency radiation**

3.3.1   Narda 8300
Use the Narda probe around and in contact with all accessible surfaces. Record highest reading obtained in each compartment of a 3×3 grid for each surface, as described in sub-section 3.4.

3.3.2   Raham 12 (82 probe)
Repeat as in sub-section 3.3.1
(a) With the plane of the probe disc parallel to the surface, and
(b) With the plane of the probe disc normal to the surface.
Also make one measurement at 30 cm from the screen surface, with the probe disc in each of three orthogonal orientations with respect to the screen.

3.3.3   IFI-EFS1
Select the most sensitive combination of antenna and scale position that is acceptable for ambient background with the VDU switched off. Then with the VDU switched on make measurements at 10 cm from the centre of each accessible surface to the centre of the antenna for each of three

orthogonal orientations. The EFS1 must be used with an auxiliary handle to minimize proximity effects between operator and instrument

3.4 **Grid system for measurements**

Where search measurements are made at the surfaces of VDU, divide the surface to be measured into an equal area 3 × 3 grid. Record the highest reading observed for each of the grid compartments. The surface monitored should be identified as follows:

Surface: Screen   Top   Base   Rear   LH Side   RH Side
Identifier: S   T   B   R   LHS   RHS

For S, R, LHS and RHS the convention for recording is 1A occupying the top left hand corner of the surface as seen by the observer. For T and B the convention is 1A occupying the remote left hand corner from the screen when the top is viewed from above and when the base is viewed from below.

APPENDIX B: Standards

X-rays

BS 415:1979   Safety requirements for mains operated electronic and
                         related apparatus for household and similar general use.
                         British Standards, 2 Park Street, London W1A 2BS.

Ultra-violet

ACGIH   TLVs—Threshold Limit Values for Physical Agents
                         1980.
                         American Conference of Governmental Industrial
                         Hygienists, PO Box 1937, Cincinnati OH 45201.
MRC   Private communication.
                         Medical Research Council, 20 Park Crescent, London
                         W1N 4AL.

Visible

Draft IEC/TC76   Technical Committee No. 76: Laser Equipment—Draft.
                         International Electrotechnical Commission, 1 rue de
                         Varembe, Geneva.
Draft BS 4803   Guide to the protection of personnel against hazards
                         from laser equipment—Draft. British Standards
                         Institute.

Infra-red

ACGIH   As above.

Microwave and radio frequency

ACGIH	As above
79-EHD-30	Safety Code 6. Ref 79-EHD-30.
	Recommended safety procedures for the installation and use of radio frequency and microwave devices in the frequency range 10 MHz–300 GHz. (Canadian) Department of National Health and Welfare, Ottawa K1A 0K9.
Draft foreign	A draft foreign national standard to be published for public comment.
MRC	Medical Research Council.
	Press Notice MRC 70/1314 dated January 1971.
	Address as above.

Chapter 6

VDU Ergonomics—A Review of the Last Two Years

TOM STEWART

Butler Cox & Partners Limited, London, UK*

VDU ERGONOMICS IS MORE IMPORTANT THAN EVER

The last two years have seen a number of developments which have made VDU ergonomics even more important. There has been major growth in the introduction of VDUs in a number of industries. They are becoming a common component not only of computer systems but also of word processing systems, cash registers, and even telephones. They are being used as a regular and routine tool in the jobs of more and more people.

A second reason for the growing importance of VDU ergonomics is the recognition that people problems are limiting the introduction and the benefits of new technology. Making the technology useful, usable and acceptable is frequently far more difficult than developing the technology itself. The limited contributions that many computer systems have made to their organizations is more often attributable to these people problems than to the technical limitations of the systems.

There is also a growing realization that the cost of adapting to bad ergonomics may not be hidden forever. More sophisticated accounting and costing techniques are allowing the real cost of errors and delays to be reckoned. But sometimes the person compensates (or even over compensates) for adverse equipment or conditions. The cost of maintained performance is therefore hidden so long as the extra effort can be tolerated. When stressful or unforeseen circumstances arise, the extra effort may be impossible to maintain. A particularly dramatic example of this occurred at Three Mile Island in the United

*Now with Systems Concepts Limited, London, UK

States. The additional stress imposed by the near disaster prevented the operators from interpreting the poorly designed displays appropriately.

The last two years have also seen significant developments in the automation of the office. The long predicted convergence of data and text processing has already started to appear in products. Integrated office systems not only result in more VDUs being introduced, they also result in far more complex operating procedures. The consequence of any errors which are made may be far more costly and significant when a number of different functions are integrated together.

There has also been a continued broadening of the interests of health and safety authorities in several countries. The emphasis has changed from concern about the need to avoid hazards and ill health to actively promoting psychological and physical wellbeing.

Public attitudes which at one time tolerated deafness and asbestosis as acceptable occupational hazards are now starting to question the acceptability of back-ache and eye strain. Working conditions which were accepted in the past are not likely to be acceptable in the future. The qualities that professional staff and managers value in their jobs (such as variety, autonomy, discretion, etc.) are now being demanded and obtained by production workers, clerks, and other lower-level staff.

The final reason why VDU ergonomics is particularly important is that we are in a period of economic recession. This means that we must get much more out of our existing equipment and people. They represent a significant investment in the past which better ergonomics would allow us to benefit from in the future. Ergonomics is seen by some as an expensive luxury which cannot be afforded when money is restricted. This is far from true. Ergonomics can be highly cost effective through reduced error rates, increased productivity, and improved staff morale and motivation.

SOME PROGRESS IN THE LAST TWO YEARS

The conference in Loughborough in 1978 concluded that VDUs can be a source of visual fatigue when they are badly designed, implemented, or used. A number of sources of discomfort and fatigue were identified and potential solutions discussed. The conference further agreed that the majority of problems experienced by VDU users could be solved by the application of existing ergonomics knowledge. However, all too often that knowledge had not been applied in practice.

Today the picture is not that different. Certainly there has been a considerable body of research throughout Europe and our understanding of the problems has now progressed. For a comprehensive statement of the current state of the art, see Grandjean and Vigliani (eds, 1980). But there are a great many terminals and workplaces which are no different from two years ago. Nonetheless there has been some progress.

VDU Design

VDU design is improving, and an increasing number of manufacturers are designing VDUs with clear stable images, adjustable screen angles, and detachable thin keyboards. The cost of terminals is falling in real terms and they are becoming a smaller part of the overall system cost. As a result designers are frequently willing to spend a little more on better VDUs with improved ergonomics.

VDU Workplaces

Workplace design shows less evidence of ergonomics being applied. They are still frequently inadequate and force the operator to adopt a cramped awkward fixed posture. But recently there has been a proliferation of special purpose VDU desks and improved office furniture. In addition posture chairs with fully adjustable back-rest seat angles are much more widely available in the UK.

Environment

Office planners and enviromental services engineers are designing more imaginative office layouts. There is a move away from unnecessarily bright offices with poorly shielded luminaires. The Chartered Institute for Building Services is providing guidance for lighting engineers specifically aimed at offices where VDUs will be used. It is becoming more common for some means of controlling the natural illumination to be provided by blinds, curtains, or solar control film.

Eye-tests For Operators

The VET advisory group has finished its work and a final report has been distributed to nearly 2000 interested enquirers world-wide. Many organizations now routinely test operators' eyesight. This provides considerable reassurance and prevents the problem of defects remaining hidden until after some period of VDU use.

Legislation and Standards

A number of organizations throughout Europe have published advisory standards for VDUs. In many cases these have come from the health and safety authorities in the country concerned, for example, the National Board of Occupational Safety and Health in Sweden and the Department of Social Affairs in the Netherlands. In West Germany much of the guidance has come from the DIN organization as advisory standards, although legislation is also being enacted to make certain aspects of VDU workplace design compulsory by 1982.

Technology Agreements

Increasingly, trades unions are negotiating technology agreements with employers before any major developments can be accepted. The majority of these agreements include health and safety considerations. Ergonomic demands are therefore becoming a major part of new technology negotiations.

More Public Awareness

During the last two years there has been growing public awareness about VDU health and safety and VDU ergonomics. Many manufacturers now emphasize the ergonomics aspects of their products. Many users and designers are arming themselves with books, articles, and research reports on VDU ergonomics in order to become demanding consumers or enlightened suppliers.

PROGRESS IS NOT WITHOUT ITS PROBLEMS

The progress described in the previous section has clearly benefited a large number of users of VDUs. However, such progress is not without its problems. Indeed, in each of the areas identified above there is also some negative progress to report.

VDU Design

Although the standard of VDU design has improved the proliferation of videotex trials around the world has encouraged the development of ultra cheap terminals. Indeed, one of the features of Videotex (also called Viewdata) is that it is able to use adapted telelvision sets as terminals for displaying text. But television was not designed for displaying stationary text; its resolution, stability, and clarity, which are quite acceptable for viewing moving pictures across a room, are frequently unacceptable for reading text in detail at a distance of a few feet. Many of the lessons learned by mistakes in the computer display field are being re-learned by Videotex display designers. Some believe that the familiarity and ubiquity of the television mean that Videotex terminals will not have to meet VDU requirements to be accepted. This seems an unwarranted and short-sighted assumption. The suitability of the terminal for the task, not its superficial familiarity, should determine its acceptability.

VDU Workplaces

Totally adjustable VDU workplaces are not necessarily an ideal. A well designed terminal with a thin detachable keyboard and the provision of swivel and tilt facilities for the screen should be able to sit on a regular desk. This may be far more suitable for the user who needs to use a terminal and to carry out tasks with conventional paperwork. Nonetheless there will be occasions where existing VDUs must be used. The adjustable desk aims to compensate for poor VDU design by providing flexibility in the desk. Unfortunately it is difficult to provide adjustable work surfaces which are both easy to adjust and satisfactory in use. Suppliers may be optimistic in what they expect an operator to do in order to adjust a work surface. On the other hand there are desks where the operator can do little more than adjust them. They are not practical for real work. Making everything adjustable is no substitute for good design.

Environment

There is one aspect of the environment which is certainly not improving with time, and that aspect is noise. Excessive noise is a major hazard in industry. Few VDU users are likely to suffer from noise induced deafness, although there can be problems in some intensive computer print rooms. However noise does not just damage hearing. At a much lower level it makes communication and concentration more difficult. The introduction of more and more office equipment—ostensibly to remove shadow functions and to free executives for productive work—is in practice making the office a less and less suitable place to work. Indeed the tasks of communicating and decision making are frequently the only activities left unautomated.

Eye-tests for VDU Operators

Although the final report by the VDU Eye Test Advisory Group has been well received, it has not been possible to set up an eye-test centre to co-ordinate the results.

Legislation and Standards

Checklists and guidelines form useful starting points, but they must be applied with caution. They must not inhibit developments in the technology. For example, recommendations about inter-character spacing typically only apply when that spacing is fixed. Standards must not be written so that proportional spacing (which is usually more legible) would be prohibited. Many current VDU typography recommendations are, in fact, restricted to current phosphors and CRT designs.

Technology Agreements

Technology agreements are negotiated between management and trades unions. This is entirely appropriate for many aspects of the introduction of new technology. However, some of the technical health and safety issues are not appropriate for negotiation. For example one negotiation resulted in a refresh rate of 57Hz being accepted as a compromise when the management had planned on 50Hz and the union had requested 60Hz. It is unlikely that such a refresh rate is either cost-effective or sensible.

More Awareness

Greater awareness of ergonomics has already led to a number of unfortunate side effects. One of these is that advertisers now use the word freely to promote their products, even where 'ergonomically designed' is totally unjustified. Second, there has been a mistaken belief that more features and more expensive terminals are inevitably better ergonomically. Ergonomic means that it suits its user and his purpose. That may mean that the simpler terminal is better. Finally, greater awareness of ergonomics has meant that many lay users and designers are referring directly to fundamental ergonomics research reports and papers. This is not a problem in itself but it does inevitably mean that the 'do it yourself' ergonomists are exposed to the uncertainties and disputes which mark the boundary of any science. When these issues are widely publicized they tend to reduce the credibility of ergonomics in areas where the facts are clearly established. There is also a danger that the latest research finding or 'health scare' is seen as the only issue of importance.

Undue emphasis on one topic can lead to far greater problems in another area. For example, placing a turntable under a terminal so that it may be rotated can make the operator's posture worse if the display is then too high. Further, the more complex and expensive the rules surrounding VDU use, the more employers will wish to limit the number of VDU users but increase their period of working. This may militate against the more successful solution which may be to make the VDU a small part of many people's job.

THE CHALLENGE FOR THE FUTURE

The previous chapters in this book have illustrated the need for more research. However, I would like to emphasize that one of the major challenges for the future is applying what we already know. The creation and use of checklists for selecting and designing VDU equipment is a useful step in the right direction. This should not, of course, be a substitute for careful consideration of the unique circumstances of each application. But even the creation of checklists does not guarantee that they will be used at the appropriate time in system development. It is therefore essential for companies to establish a formal ergonomic review stage in their system development programmes. This review should take place before systems progress into implementation. It should act as a formal check and prevent unworkable systems being imposed on users.

The review should not just be concerned with hardware. In many cases the software of the system (the dialogue) virtually defines the user's job. Yet the software designer was often unaware that he was, in fact, designing a job. It is not surprising, therefore, that many computer based jobs are neither rewarding nor satisfying. Poor software design may be far more expensive than poor hardware and is often much more difficult to change. Getting the software interface correct from the start is therefore one of the major challenges for system designers in the coming years.

CONCLUSION

VDU ergonomics has progressed in the last two years. Our knowledge of the issues has expanded and new problems have been identified. Yet our ability to apply our knowledge has not developed apace. Our efforts in the near future should be concentrated on eliminating the unnecessary problems with hardware and making a real start on improving VDU software ergonomics.

REFERENCE

Grandjean, E. and Vigiliani, E. (eds) 1980. *Ergonomic Aspects of Visual Display Terminals*, Taylor & Francis, London.

Chapter 7

Cataracts and Visual Display Units

MILTON M. ZARET, MD

Director of Research,
Zaret Foundation, 1230 Post Road, Scarsdale, NY, USA

INTRODUCTION

In furthering the aims of this book it may be of value to provide some background information concerning the origins of the VDU cataract problem. Following this, material will be presented concerning two recently diagnosed cases, each of which adds an additional, different dimension to the problem. Then, as perceived in the context of the cataract problem, it may be possible to define 'responsible and rational' in a meaningful way.

BACKGROUND

While the United States remained mostly unconcerned, European scientists, primarily ergonomists, made significant progress towards both discovering and, more importantly, correcting many VDU design defects. I first became aware of this from an excellent paper by Östberg (1975) of Sweden and was impressed by the prominence of asthenopic and neurasthenic signs and symptoms in some VDU operators. Although ergonomic dysfunction could explain most of the findings, it soon became evident to me, based upon my knowledge about the biological effects of non-ionizing radiation exposure and the associated clinical manifestations of asthenopia and neurasthenia, that ergonomics may not be the only factor to be considered.

Prolonged exposure to some non-ionizing radiations can lead initially to the subtle development of a neuroasthenic-like syndrome which, at this stage, appears to be physiological stress and reversible. However, if repeated exposure continues, that leads ultimately to an irreversible, pathological state. This was

47

reported by Sadcikova (1974), who tabulated the findings of a hundred cases of microwave or radiowave sickness which has long been recognized as an occupational disease in the Soviet Union. Some of her cases, exposed to field intensities in the vicinity of 1 mW cm^{-2}, also developed cataracts.

Not only Sadcikova but also Hirsch and Parker (1952), Carpenter and Donaldson (1970), and Zaret (1974) all reported cataracts in men following repeated exposures to non-ionizing radiation at energy densities within an order of magnitude of 1.0 mW cm^{-2}.

That electromagnetic radiation could produce cataracts has been known for more than a century. Indeed, as long ago as 1926, Sir Stewart Duke-Elder (1926) in a brilliantly prescient manuscript about cataractogenesis showed that cumulative exposure to radiant energy, including long-term irradiation by sunlight, was the primary sensitizing etiological factor for most types of cataract acquired during life. The data Sir Stewart presented indicated that rays originating anywhere throughout the electromagnetic spectrum, from the longest wavelengths generated by electrical oscillations to the shortest wavelengths of ionizing radiation, had the property of initiating cataract formation.

An objective sign having important differential diagnostic significance appears to be the acquisition of capsular opacification at the posterior surface of the lens (recognized as the refringent edge of the lens when viewed by slit-lamp biomicroscopy). When this finding makes its initial appearance after repeated exposure to non-ionizing radiation in the absence of any other significant cataractogenic factor the finding of capsular cataract can be considered to be pathognomonic for radiant energy injury. Regarding the invisible portion of the non-ionizing radiation spectrum this was first described by me in 1964 (Zaret, 1964) in a microwave technician and then verified independently in 1967 by Bouchat and Marsol (1967) of France. They reported the case of a radar technician whose eyes were normal when he began work at the age of nineteen. Over the next four years they observed him to develop capsular cataracts, ruled out by differential diagnosis any other known causal factor except for microwave irradiation, and, therefore, published their findings so as to alert French ophthalmologists to the diagnostic relationship of microwave exposure to capsular cataract.

Radiant energy can also produce other forms of cataract directly or can be an additional or co-factor in accelerating still other types of cataract. However, acquired capsular cataract, as defined above, served as the criterion response for the eight cases of cataracts following use of cathode ray tube visual display systems that I reported in 1980 at the International Symposium of Electromagnetic Waves and Biology (Zaret, 1980) as well as for the two new cases being reported here.

My involvement with VDUs, VDTs, and CRTs began in 1977, but not from a vacuum. I initially worked on problems of microwave and radio frequency radiation. This led to work with long infra-red and to the visible portion of the spectrum with near infra-red and ultra-violet. The latest work involved ELF in the opposite direction, so we have now covered what is termed the non-ionizing radiation spectrum, concerning ourselves with the potential problems.

With human beings it is very difficult to retrospectively record all the exposures they may have had to various radiations, but we try to achieve this through the process of 'documentation'. One thing we are able to document, by photographic techniques, are certain characteristic lens injuries which may lead to cataract formation.

The first microwave cataract I saw was in a man whose job was that of a microwave tube technician. To test prototype microwave tubes the man had to look through a window cut in the casing of the tube and adjust the current until the circuit energized properly. When one eye started to fail, he used the other eye. The plant's optometrist referred him to other ophthalmologists. All agreed that he had a radiation cataract of some sort but were unable to find any significant levels of radiation at his workplace. He continued with his job until eventually I saw him. Examination with the aid of a slit-lamp biomicroscope revealed that opacification had occurred on the posterior capsular surface of the lens.

Other factors can cause opacification, other than radiation. However, this man had no history of other causes, as was discovered by differential diagnosis. This injury differed from the other common type of capsular cataract called 'glass blowers' cataract, which has now virtually disappeared. The British Home Office, in 1905, made 'glass blowers' cataract a compensatable disease, and the introduction of effective protective goggles in this industry was found to be cheaper than paying compensation. Glass blower's cataract frequently affects the anterior capsule first; whereas radiowave cataract usually starts at the posterior capsule.

A posterior capsular cataract was also found in a former radar officer, which developed many years after his exposure. The effects of radiation can continue for many years. (This type of cataract was commonly seen in the UK before 1905.) Other changes had also occurred in his lens. He had nuclear sclerosis, where the central portion of the lens also becomes opaque, which can lead to an opacity of the entire lens. The posterior capsular cataract was very extensive with many shapes and configurations, but the posterior sub-capsular region was clear of cataract. In this case, nuclear sclerosis was a sign of premature abnormal aging—another known effect of radiation.

Another case deals with a man who was examined periodically over many years at the Atlantic Missile Range. This facility had excellent radiation monitoring systems and personnel who were very competent in taking measurements. This man had never been exposed to 10 mW cm^{-2}, which was the standard permissible level of radiation at that time. Nevertheless, after a number of years working in the field of less than 10 mW cm^{-2} microwave radiation he developed a cataract; an opacification on the posterior surface of the lens.

Radiation cataracts are not only caused by consumer products and radar sets; a physician developed a cataract in the course of his work. He used microwave diathermy extensively in his office for a number of years. He was never treated by diathermy but was present with his patients very often. With this as a background, the rest of the cases are all related to visual display units.

REVIEW OF EARLIER PAPER (Zaret, 1980)

In 1977, two young newspapermen, Case 1 aged thirty-four and Case 2 aged twenty-nine, were each referred to me separately for ophthalmic consultation because each had acquired incipient cataracts shortly after beginning work with VDUs. For both of them, all the other etiologies for acquired cataracts except for radiant energy injury were excluded by differential diagnosis. My examination revealed that the diagnosis for both men was radiant energy cataract, the features of which implied that the only ratiocinate site of exposure was at the *New York Times*, where both of them worked, with recently installed cathode ray tube visual display systems, as copy editors.

In that context, these two cases represented an unusual cluster because they had multiple interrelationships of epidemiological significance; i.e. a signature of radiant energy injury in addition to both spatial and temporal proximity to the same enviroment! This set of facts did not, *per se*, prove that the occupational enviroment at the *New York Times* was at fault. Instead, it indicated that priority should be given to investigate that enviroment properly and thoroughly with special attention being paid to the CRT generated displays and their associated equipment, but that, in addition, all other potential sources of exposure at the newspaper's offices should also be investigated.

At this juncture the *New York Times* requested that its consulting ophthalmologist examine the patients and my findings were provided to him. He found that the acquired cataracts in both patients were compatible with radiant energy injury. Moreover, as I had also done previously, he too, effectively excluded, by differential diagnosis, all other ordinarily encountered factors for the cataracts.

The determinant question then became not the moot point, whether radiant energy probably caused the cataracts, but, instead, whether or not the occupational environment at the *New York Times* could have contained harmful spurious non-ionizing radiations? Unfortunately, the newspaper, the equipment manufacturers, the responsible governmental agencies and, eventually, even an arbitration tribunal all examined the moot point, instead of the germane issue. And, in error, they all concurred there was no hazard in spite of the fact that the US National Institute of Occupational Safety and Health (NIOSH) had discovered a level of irradiation of 1 mW cm^{-2} in the *New York Times*' building!

Their collective error in dismissing VDUs as being causally related was confirmed in 1979 after I discovered Case 3, a newspaper woman aged forty-nine, who developed radiant energy cataracts after working with VDUs for a duration of six months. In analysing the then current literature it became obvious that the arbitrary standard NIOSH applied for ultra-violet radiation safety was just as wrong as its arbitrary standard for radio frequency radiation had been, and that normally operating VDUs could emit cataractogenic levels of UV radiation (another region of the non-ionizing radiation spectrum) as well as invisible non-ionizing radiations.

The above finding of a cluster of cases, each fulfilling a relatively stringent criterion for the clinical diagnosis, radiant energy cataracts, and each acquired only after intimate contact with cathode ray tube display systems was sufficient evidence to suspect that a real problem existed. Nevertheless, accumulating meaningful human data about radiational injuries is extremely difficult. Indeed, to this date, it has never been accomplished in a completely satisfactory manner. One obvious reason is that humans do not live *in vacuo*; instead, they are bathed by a normally occurring as well as an artificially created radiational environment and both are variable. In this context, if one also considers that the effects of radiation are both additive and cumulative (neither in an exact mathematical relationship!), then this complicates matters further.

Despite those problems, by utilizing the clinical constraints of differential diagnosis in some cases the additive and cumulative effects of radiation can be recognized with relative confidence. This is demonstrated by the following two cases, where their own ophthalmologists documented the fact that an active process of lens opacification made its clinical appearance only after working with VDUs.

Case 4, a radar technician aged forty-one, exhibited the earliest evidence of microwave radiation injury, roughening and thickening of the posterior capsule in both lenses, when first examined by me in 1960. By 1969 there was a marked asymmetry between the two eyes, a radiant energy (microwave) cataract had formed in his right eye while his left lens exhibited only a very early stage of capsular opacification, and he was one of the forty-two cases of microwave cataract reported by me that year (Zaret, 1969).

The patient was then transferred to office work where he had no further substantive exposure to microwave radiation and the cataractogenic process in the left eye became arrested in a forme fruste state. It remained quiescent until about 1976 when it then began to show definite signs of progression and then the cataractogenesis accelerated so that by 1978 surgery was required to restore vision.

This was an unexpected course because, when removed from further exposure, it is unusual for a microwave injured lens arrested in a forme fruste stage to undergo the type of reactivation as occurred here. Further, the patient did not have any known sources of exposure away from work such as from a microwave oven or CB radio transmitter. In obtaining a retrospective work environment history it was discovered that he had begun working daily with a VDU about six months prior to the reactivation of the cataractous changes and therefore this case can be viewed as representing the cumulative effect of non-ionizing radiation.

Case 5, an office worker aged fifty-three, exhibited the additive effects of non-ionizing radiation to a prior sensitization by ionizing radiation. She was first examined by me in 1980 because of unilateral cataract that recently appeared in her right eye. Differential diagnosis by her treating ophthalmologist and also by another consulting ophthalmologist had ruled out all other known causes for cataracts except for radiant energy exposure. Although both her lenses exhibited an early stage of capsular opacification, other features of the slit-lamp findings in her right eye indicated that it had been exposed to radiant energy a long time ago, prior to the four years she had been working with VDUs. So I obtained copies of all prior medical records and discovered that she was, in fact, exposed to ionizing radiation when she was sixteen years old at which time she had X-ray therapy of her face for acne (an accepted form of therapy then). This indicated that her recent exposure to VDU radiation was a contributory etiological factor for her unilateral cataract, that prior X-ray exposure had sensitized the lens, and that the cataract was due to the additive effect of the two different types of radiation.

Radiant energy cataract can evolve concomitantly with other pathologies and there is much in the literature which indicates that injury can occur to neurological function and special sense organs, like the ear, following repeated exposure to non-ionizing radiation. One report chronicles the evolution of blindness, deafness, and vestibular dysfunction in a microwave worker (Zaret, 1975).

Case 6, a female office worker ageed forty-four, was referred to me early in 1980 because she had acquired an uniocular incipient cataract after working with VDUs for three years. Late in 1978, after working with VDUs for eighteen months, a bizarre illness started with a partial loss of hearing in one ear associated with tinnitus, vestibular dysfunction, and a loss of pupillary reactions. Extensive clinical and laboratory testing, including neurological, otological, and ophthalmological consultations, failed to establish any etiological diagnosis for any of her pathological findings until the incipient radiant energy cataract was found in her left eye (also, her right eye exhibited the earliest signs of non-ionizing radiation injury).

Another aspect to the VDU cataract problem concerns the special vulnerability of some individuals who may have a serious pre-existing condition. This was exemplified by Case 7, a male aged thirty-three, examined in May 1980 because the patient had only one eye. It had recently acquired a cataract and the cataract was suspected as being due to his work with VDUs.

This patient had been born prematurely and unfortunately acquired retrolental fibroplasia (RFL) which led to blindness of the left eye and partial retinal degeneration of the right eye. Prophylactic photocoagulation to prevent an impending retinal detachment was performed in his right eye in 1967. In addition, being highly myopic, he brought his face very close to an object in order to see it clearly. Thus, when he worked with VDUs constantly from 1977 to 1979 and then intermittently afterwards his face was within 3 to 4 inches of the surface of the tube for viewing. Shortly thereafter his vision began to fail, his ophthalmologist determined that a cataract was forming, and the patient was referred to me for consultation because it was suspected that VDU radiation could have initiated this cataract.

There are certain industries, like aviation, where visual disturbance in an individual worker can create the risk of injury or even death for others. For example, the following case of radiant energy cataracts in an air traffic controller was reported first in relationship to avionic environments (Zaret and Snyder, 1977) and, later, in relationship to air safety (Zaret, 1979). At those times it was not possible to identify a single source of exposure as the predominant causal agent because air traffic controllers can work near many types of radiating equipment, such as radar sets, radio transmitters, microwave telecommunication systems, and radio navigation devices as well as radarscopes.

In this case, however, the Federal Aviation Agency (FAA) determined that the patient had no significant exposure to any of its high power hertzian radiation transmitters because his principal work was only associated with viewing radarscopes. As most of us now know, the main element of a radarscope is a cathode ray tube similar, in principle, to that of a VDU.

The patient (Case 8), a male air traffic controller aged fifty-four, was examined by me in 1976 because his attending ophthalmologist believed that the cause of his cataracts was due to prolonged viewing of radarscopes. I confirmed that the patient exhibited a mature stage of non-ionizing radiation cataract in his left eye and an incipient stage of non-ionizing cataract in his right eye. However, at that time I did not have sufficient information to determine, with confidence, which radiating equipment was at fault so I did not render an opinion then as to which of the various possible equipments caused the exposures.

His history was quite interesting. For twenty-four years, from 1948 to 1972, practically all of his work was performed by viewing radarscopes. His eyes and vision were normal until 1967, when intermittently he began to experience visual and ergonomic complaints. Occasionally he made errors in identifying aircraft on his radarscope and placed them into potential mid-air collision courses. It was not until 1970, at the age of forty-eight, that an incipient cataract first appeared in his left eye. Nevertheless, even this was continuously waived by the FAA despite repeated episodes of 'losing' aircraft, until 1972 when he caused near misses on two successive shifts.

Subsequently, the patient filed a claim for work related injury. The FAA denied responsibility based on its determination that the patient had no significant exposure to any of its high power hertzian transmitters (apparently it considered that only high power sources could cause injury). The FAA did substantiate the patient's claim that his principal work was associated only with viewing radarscopes. Now that we know that cathode ray tube viewing devices can emit cataractogenic levels of radiant energy it is reasonable to concur with the attending ophthalmologist's original opinion that viewing radarscopes, in particular, was the cause of the cataracts, because now this can be rationalized with confidence.

In summary, the cases selected for presentation represented examples which demonstrated different facets of the overall problem. Cases 1 and 2 were a cluster implicating the occupational environment of the *New York Times*. Case 3 indicated that the problem was not limited to the *New York Times* but also could occur at any other newspaper with a similar installation. Case 4 was an example of the cumulative effect of non-ionizing radiation; whereas Case 5 was an example of the additive effect of non-ionizing radiation to prior exposure by ionizing radiation. Case 6 indicated that other organ systems, in addition to the eye, could be affected adversely. Case 7 showed that individual vulnerability must be factored into any equation purportedly representing safety for the individual. And Case 8 demonstrated that, for some critical occupations such as avionics, reduced cognition in a CRT operator could result in injury to others.

NEW CASES

In order to appreciate better both the nature of the VDU problem and the difficulties encountered in its resolution some salient features of two newly discovered cases will be presented.

Case 9, a female computer programmer aged forty-two, was referred to me in September 1980. She had been working full-time with a VDU for the past five years; she first exhibited symptoms of asthenopia one year ago and these have recurred intermittently with increased frequency and intensity. She noticed a gradual, aberrant, difference of colour sense between her two eyes (anisometachromatopsia) and she stated that her symptoms were aggravated only when she worked with the VDU screen but not when she read print-out or other written material. The salient findings of the ophthalmic examination were a weakness of convergence and early presbyopia, both contributing to the ergonomic dysfunction and a small area of radiant energy injury in the peripheral 10:30 o'clock meridian at the posterior capsular surface of the right lens. As in a small but growing number of similar cases, here there appeared to be an interrelationship between or a merging of the visual, ergonomic and cataractogenic factors associated with VDU generated physiological and pathological disorders.

Case 10, a male air traffic controller aged twenty-seven, was examined by me in August 1980, having been referred by his ophthalmologist because the ophthalmologist had a very disturbing experience with the FAA's management representative. It epitomizes many of the difficulties encountered by physicians when they attempt to deal with the VDU health problem after the setting has become adversarial.

The patient had been examined by his ophthalmologist many times between 1969 and 1977 and the intra-ocular tissues, including the crystalline lenses, were found to be normal at every examination. However, the patient noticed blurred vision of his right eye which persisted for a few days, and this was the chief complaint leading to the ophthalmological examination in July 1980. The pertinent findings were that, although visual acuity was correctable to 20/20 for the right eye and 20/20 for the left eye, nevertheless, the patient had acquired posterior 'subscapsular' opacities bilaterally, more in the right lens than in the left. As the patient was otherwise in excellent health, all other known causes for the incipient cataracts had been eliminated, the cataracts were of a type that could be produced by exposure to radiant energy, and he had been employed as an air traffic controller since 1977, his ophthalmologist attributed the cataracts to occupational exposure to radiation, and he so notified the FAA physician.

The FAA physician stated that no microwave emanations were generated at the facility where the patient worked. The term 'microwave' was used here loosely in accordance with general knowledge: a more correct term, technically, would have been 'non-ionizing radiation' or 'radiant energy' to indicate the electronic smog aspect of the occupational environment at FAA installations. In this context, it was grossly inaccurate and misleading to imply that no such emanations are present at the facility because electronic processing is present at all air traffic controller facilities. As if that were not bad enough, the FAA physician implied that the cataracts were congenital, and, therefore, had been present since birth, despite not only the patient's private ophthalmologist having documented that this was not so but also that the FAA's annual medical examination records for 1977, 1978, and 1979 also documented that this was not so. Instead, all these records indicated clearly that the cataracts were first acquired when the patient was twenty-seven years old and they were not present at birth. In addition, the FAA physician made a note in the patient's medical record that the FAA would refer the patient to its consulting ophthalmologist because the case represented a 'possible adverse action'. The FAA's consulting ophthalmologist examined the patient early in August 1980 and, as anticipated by the FAA physician, diagnosed the condition as congenital cataracts.

At this juncture, the patient's ophthalmologist referred the patient to me for examination and opinion. The patient exhibited posterior capsular cataract, bilaterally, with the right eye being more advanced than the left, and these findings were pathognomonic for non-ionizing radiant energy injury in this case. My diagnosis was subacute incipient radiant energy cataracts.

In view of the unorthodox and apparently unwarranted diagnosis put forward first by the FAA physician and then by his consultant I requested and obtained copies of the actual medical records before preparing my report. There were two basic reasons for this. The first had to do with preventive medicine, because the patient could possibly avoid or, at a minimum, delay the cataract from progressing to a stage where surgery would become necessary by removing him from the harmful occupational environment. The second reason had to do with determining whether or not there existed any factual basis for the diagnosis of congenital cataracts in this case. No credible rationale existed, which brings into focus a point frequently overlooked by individuals having parochial interests in cases such as this: no matter how adversarial the employee–management equation may become there should not be any room whatsoever for the practice of adversarial medicine.

In that context it is interesting to note that the FAA had taken a position that the patient was not exposed to any significant radiation from its high power radio-frequency transmitters because the patient worked only with cathode ray tube visual display systems. The patient confirmed that all of his work was with cathode ray tube visual display systems. I, too, accept that as being a medically reasonable fact in this case. Indeed, that serves as the basis for including this case in this chapter.

The emanations from visual display terminals are also the cause of other problems. 'Asthenopia' is an ophthalmological term which relates to problems of ocular fatigue and distress which arise from an imbalance of the musculature in the two eyes working as a team. 'Neuroasthenia' has a different group of symptoms and may arise from asthenopia. However, we have seen neuroasthenia in patients exposed to radiant energy but who do not have muscular imbalances.

The American standards (proposed and implied), used for non-ionizing radiation of the invisible portion of the spectrum, RF-microwave radiation, radiowave, and hertzian radiation are all arbitrary. With the *New York Times* dispute, the management and manufacturers of the equipment denied that there was any radiant energy leaking from any of the devices. However, the eyes indicated that there was. I, and other physicians, have found radiant-energy cataracts in humans who had been exposed to radiation of only $1 \, \mathrm{mW \, cm^{-2}}$ which is less than the 'acceptable', permissible standard levels.

The evidence of cataracts does not prove that they were caused by the VDUs, but they were caused by something. The medical reasonableness of the situation is convincing enough, however, to make it the only ratiocinate reason for the patients to exhibit these clinical stages of cataract. The arbitrator's consultants in the *New York Times* case apparently used the approach that there was meaning to the standards, that the standards were absolute. Their argument followed that if the standards were not exceeded then there could be no causal relationship. That was an error.

That situation lay fallow until two years later, when an unrelated trade union insisted that measurements be made at their VDU work stations. They found levels in excess of the 'microwave' standard coming from the operational units. This then led to a re-evaluation of the situation at the *New York Times*, and they found that there were, in fact, levels of irradiation which exceeded the permissible standard utilized by the arbitration. That standard permitted a level of 10 mW cm^{-2}, but it had no time-related boundary. For continuing exposure one should be exposed to only 1 mW cm^{-2}. For one tenth of an hour per hour this could rise to 10 mW cm^{-2}, giving an average power of only 1 mW cm^{-2}. That is the pragmatic application of the contrived standard.

A correction was also supposed to be made for the temperature and humidity index, involving a complicated formula. When there is a rise above 70°F but below 79°F there needs to be a reduction in the permissible radiation levels. Ultimately, for a temperature/humidity index over 79°F, only a tenth of 1 mW is permitted. That is 100 μW cm^{-2}. However, none of the measuring equipment used in any of the surveys were sensitive to that level, as far as I am aware. In addition, the American standards were set for normal healthy adults. With certain sicknesses (which were ill defined), such as circulatory problems, the radiation levels were to be further reduced by a factor of ten. That was how this 'technology' was applied in those days when the radio frequency standard was evolving: a prudent physician could be expected to apply that factor for the unknown.

This, then, reduced the American standard to 10 μWcm^{-2}, which is precisely the same as the so called 'Russian' standard. Under certain circumstances that can be exceeded up to 100 μWcm^{-2}. This may apply to soldiers or radar technicians who, for defence reasons, have to make adjustments to equipment which is still functioning. For brief periods of time, the Russian worker could be exposed to 1 mW cm^{-2}.

The American and Russian standards are essentially the same, although the approaches differ. They are both wrong. Unfortunately the standard setting committees are not the people who will have to apply these standards; instead, people like us will have to. It should be understood that, as a general rule, the more dogmatic and firm people are in their statements about radiation safety, the less they really know about the subject.

The radiations emitted from the VDU include not only infra-red and the invisible non-ionizing radiations but they also include ultra-violet. This was apparent from certain characteristics seen in the eyes of some patients. Some VDUs have already been found to be leaking at the frequency of the flyback transformer (about 27 MHz) in excess of the recognized (30 MHz) standard. There are other areas which need investigation. One example concerns the sweep speeds of about 15–16 kHz: if there are malfunctions in those circuits, radio waves can occur. (Note: subsequent to this writing, many VDUs have been found to emit kilohertz radiation.) If parasitic oscillations occur in the CRTs energy can build up on the electron gun. The configuration of the gun makes it a potential transmitter of radiation in the gigahertz region. A marriage can also occur with electric typewriters so that in some areas ELF can be created.

Other areas have yet to be explored. We do not know the causes of all of the radiant energy problems, therefore it will be some time before it is known how

VDUs can be made safer. This does not imply that most VDUs are not already quite safe.

Concerning the subject of ultra-violet radiation, the permissible level stands at 1 mW cm^{-2} in the area between 330 and 400 nm. A dental device has been developed which uses ultra-violet radiation to harden plastic fillings *in situ* in the the tooth. Quite a number of dentists and dental technicians working with these devices have now started to develop radiant energy cataracts. Dr Sidney Lerman, an American ophthalmologist, was called in to investigate the problem. He discovered that about 1 μW cm^{-2} can be cataractogenic for some sensitive people. That is a factor of a thousand less, three orders of magnitude less than that considered to be a safe level of ultra-violet radiation by some authorities in the United States.

Further research into the causes of radiant energy cataracts is hampered by the lack of ophthalmologists trained to look at this problem. In addition there is the difficulty of gauging an individual's susceptibility to cataracts. Finally, there is the added complication of trying to establish the past history of radiation exposures that have been encountered by the patient.

SUMMARY

Without question, the major issue before us is how to resolve VDU health hazards in a responsible and rational manner. The method to accomplish this is simple in theory but difficult in practice. The first step is to define not only 'hazards' but also 'responsible' and 'rational', because these terms are perceived differently by vested interest groups. The next step is to implement meaningful research programmes and to obtain the services of the best scientists, instead of the best that money can buy, in order to resolve the first tier of important questions. Otherwise, the prime need, the establishment of meaningful health and safety procedures to protect VDU users, will become obfuscated by the adversarial issues which are of lesser importance than preventing radiational injury.

REFERENCES

Bouchat, J., and Marsol, C. (1967). 'Cataracte capsulaire bilaterale et radar', *Archives d'Ophthalmologue de Paris*, **27**, 593.

Carpenter, R.L., and Donaldson, D.D. (1970). 'Bilateral cataracts following microwave diathermy treatments: a case history', *5th International Symposium*, International Microwave Power Institute, Scheveningen, The Netherlands, p. 14.

Duke-Elder, S. (1926). 'The pathological action (of radiant energy) upon the lens, *Lancet*, **1**, 1188.

Hirsch, F.G., and Parker, J.T. (1952). 'Bilateral lenticular opacities occurring in a technician operating a microwave generator', *AMA Archives of Industrial Hygiene*, **6**, 512.

Östberg, O. (1975). 'CRTs pose health problems for operators', *International Journal of Health and Safety* (Nov./Dec.), 24.

Sadcikova, M.N. (1974). 'Clinical manifestations of reactions to microwave irradiation in various occupational groups', in *Biological Effects and Health Hazards of Microwave Radiation*, ed. by P. Czerske, Polish Medical Publishers, Warsaw, p. 261.

Zaret, M.M. (1964). 'An experimental study of the cataractogenic effects of microwave radiation', *Technical Documentary Report, Contract AF 30(602)-3087*, Griffiss Air Force Base, New York, RADC-TDR-64-73:13.

Zaret, M.M. (1969). 'Ophthalmic hazards of microwave and laser environments', *Proceedings of Aerospace Medical Association Annual Meeting*, p. 219.

Zaret, M.M. (1974). 'Selected cases of microwave cataract in man associated with concomitant annotated pathologies', in *Biological Effects and Health Hazards of Microwave Radiation*, ed. by P. Czerski, p. 294.

Zaret, M.M. (1975). 'Blindness, deafness and vestibular dysfunction in a microwave worker'. *The Eye, Ear, Nose and Throat Monthly*, **54**, p. 291.

Zaret, M.M., and Snyder, W.Z. (1977). 'Cataracts and avionic radiations', *British Journal of Ophthalmology*, **61**, p. 380.

Zaret, M.M. (1979). 'Air safety', *New York State Journal of Medicine*, **79**, p. 1964.

Zaret, M.M. (1980). 'Cataracts following use of cathode ray tube displays', *Proceedings of International Symposium of Electromagnetic Waves and Biology*, Jouey-en-Josas, France.

Chapter 8

Summary

PROFESSOR GRANDJEAN

Director, Swiss Federal Institute of Technology,
Zurich, Switzerland

I will start with Brian Pearce's chapter which explained initially why the conference was organized. He believes that to discuss reports of health injuries as being of little consequence is irresponsible. Many speculative assumptions about health injuries are possible in this field and the question arises as to when does a rumour of a health hazard merit or dictate investigation. He emphasized the importance of the on-screen exposure time and is in favour of a reduction to 4 hours daily. This will no doubt be a point of discussion. Furthermore, Brian suggested the setting up of an agency to be concerned solely with the human aspects of new technology, which could respond quickly to allegations of health hazards.

This chapter was followed by two others dealing with a similar topic; the facial rashes and red-skin reactions among VDU operators. Dr Rycroft described four cases; of course, with only four cases one cannot make a statistic, but all appreciated that he was careful enough not to draw any dangerous conclusions from them.

Dr Tjønn, from Norway, then reported about forty cases, sixteen of which had all the symptoms of dermatitis. He mentioned the 'provocation' tests which were performed in clinics. Here the operators had only one part of their face exposed to the screen but there were no reactions. When the same test was performed in the workplace the provocation tests gave positive results. This indicates that it is not the visual display terminal by itself which causes the problem but that there is another factor present in these work rooms. Carpets were thought to be a possible factor. Both authors mentioned the presence of a great deal of static electricity. Dr Tjønn referred to some experiments with static and came to the conclusion that some VDU operators may acquire contact dermatitis from airborne dust influenced by the static.

This leads on to the chapter by Mr Cox. Obviously this static electricity has something to do with the electrical field he measured. In his report we see that a certain number of VDUs had fields from 10 to more than 200 Vm^{-1}, but omits to mention the threshold limit value. If a threshold limit has been established it is probable that it was not assessed in the light of the VDU. Mr Cox covered the whole field of radiation emissions and in every case found the emissions to be far below the usually accepted and assessed threshold limit values.

Tom Stewart has shown that within the last few years some progress has been made in the field of VDU design. On the other hand he has also shown that progress has its negative aspects, with examples of poor ergonomic designs in the cheap television terminals on the market. He was also unsure whether the adjustable work stations available constitute progress or not. In many cases the software of the system defines the user's job and therefore some of the pressures. He closed with a sentence stating that there is still a big gap between what we know and what we do. I agree with this statement.

I would like to illustrate something that Tom Stewart said. We did an evaluation of eight different VDTs using all sorts of criteria, and Figures 1 and 2 show the great differences in the ergonomic quality of two of them. Tom Stewart said that the characters should have sharp corners and be steady. We can measure the luminance of one dot. If the dot is steady the luminance is constant; if it is moving the luminance will move also. A good terminal will have a very steady luminance (A, B, C, and E in Figure 3). A bad terminal (D, F, E, and H in Figure 3) with a very unstable luminance will certainly be unsuitable for the accommodation of the eye.

Figure 1. The 'legs' of the letters are short and have good contrasts with their near-vicinity. The letters have a good relationship between height and width, and the spaces between letters and lines are large

Figure 2. The spaces between letters are not constant and the 'legs' are often merged. This phenomenon is, to some extent, due to the instability of characters. The whole face is poor

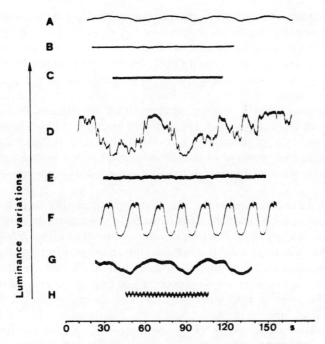

Figure 3. The stability of characters of the eights VDTs. The luminances of dots (0.08 mm) in the middle of a 'leg' of one character are recorded. A poor stability is reflected by an increased variation in luminances

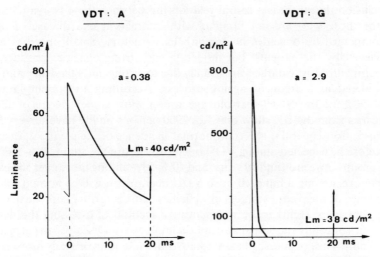

Figure 4. Luminance oscillation of dots (0.08 mm) of two VDTs with approximately the same mean luminance. Lm = mean luminance

We also studied the oscillating characteristics. The light of the characters is not a constant one as under natural conditions, but with each electron beam passing there is an illumination of the dots. We measured the oscillation of the very small dots of the characters. In Figure 4 (VDT: G) the light intensity rises to its peak value then falls to nearly 0%: between two oscillations there is practically no light. The human brain does not follow that and does not distinguish the oscillating peaks, but it is most probable that the retina will follow these luminance changes. Figure 4 (VDT: A) is an example of a much better VDU, where the oscillating luminance only drops to about 25% and goes down slowly. It is probable that this type of luminance is much better for the retina.

I wish to emphasize the great differences in the quality of the terminals on the market.

I would like to close this part of the summary with another illustration (Figure 5). The black areas indicate the parts of the body where more than 10% of people complained of daily pain. The four illustrations represent four different groups of workers. Reading from the left are the data entry terminal group, the conversational terminal group, the typist, and finally the traditional office worker. One can now see why it is necessary to limit the time spent on VDUs. I would also propose that work organization should involve a mixture of conversational and data entry operation. Furthermore, if the ergonomic recommendations concerning VDTs and work stations are taken seriously one can then return to the normal incidence of impairment, as illustrated in the diagram for the traditional office worker.

Now we come to the last part of the summary, to the chapter by Dr Zaret.

Dr Zaret showed us ten case histories of cataract patients. All ten cases, seen in the last four years, had worked with VDUs or CRTs. Dr Zaret draws the conclusion that there is a causal relationship between these two situations.

For those who are not familiar with cataracts it is a disease of the lens accompanied by opacities in its epithelial structure. Small opacities are not perceived by the patients, but moderate and strong opacities produce visual impairment. Most cataracts are usually due to a congenital tendency and can be considered as a frequent ageing process. According to present knowledge 40%–50% of the 50–60-year-old age group show some opacity of the lens. Statistics from the UK show that 1 in 2000 persons finally have surgery for this disease. Consequently, a certain normal incidence of incipient cataracts must therefore be expected among VDU operators. I estimate about 20%–30% in the age group between 30 and 50 years, and 40%–50% among those over 50 years old. Therefore, among a hundred or so VDU operators it will be normal to find an incidence of incipient cataracts or opacities. This is further complicated by the fact that there seems to be no quantitive method of assessing the degree of opacities or cataracts, perhaps an ophthalmologist or Dr Zaret can correct me on this point. This, of course, makes a statistical evaluation or study more difficult.

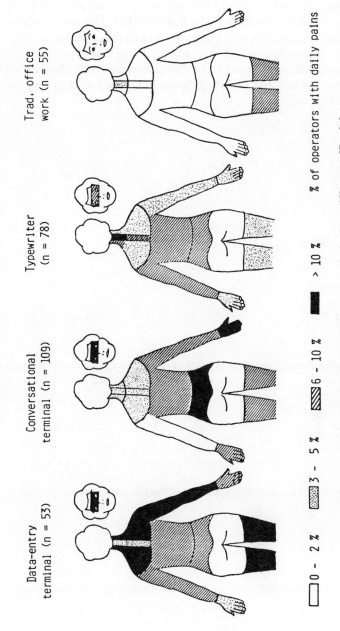

Figure 5. Incidence of daily bodily and eye pain in four different office jobs

It is true that some radiant energy of the electromagnetic spectrum is cataractogenic for human beings. Microwaves have been found to produce cataracts, as has infra-red, as Dr Zaret mentioned in reference to glass blowers. Here it is considered to have a local thermal effect. The amount of energy needed to produce a cataract is very high and fortunately in the human situation, rare. However, many experiments have been performed on animals, and microwaves and ultra-violet were found to be cataractogenic. Only one part of the waves penetrates the lens and the whole X-ray radiation and corpus colour radiation can produce cataracts.

Chapter 9

General Discussion

A national official of a trade union asked Brian Pearce whether he was aware of a problem in Grimsby in 1977–78 involving five women; she thought that three had had miscarriages, one a still-birth, and another a deformed baby. Brian Pearce explained that he was not aware of any scientific study that had investigated that report. It was suggested by the trade union official that the Health & Safety Executive had investigated that report and that they may be able to give further details.

Dr Rycroft was asked what happens in the skin with facial rashes. He emphasized that he and his colleagues had not seen the rashes, so they were at a disadvantage in interpreting the appearance. The appearances on the face were rather non-specific, in other words, they could be derived in many different ways. Redness, scaling, or spottiness on the skin may be caused by an irritant, an allergen, or from drying and dehydration.

An occupational hygienist from a computer supplier who was involved in the incident which Dr Rycroft reported, thought that there was considerable uncertainty about the humidity level figures mentioned. One individual did have a very sensitive skin and experienced reddening when he was exposed to sunlight. The other was extremely worried about harmful rays from the VDU. It had been suggested that skin rashes may possibly be of psychosomatic origin, this may have been a factor in his particular case. On the question of looking at the screen from an oblique angle he recalled that the individual was not, in fact, any further away from the the screen after it was altered. As the ultra-violet would have been emanating from the phosphor it seemed possible that he may still have been receiving the same dose had he been looking straight at the screen. It was a dangerous assumption to point to a single factor, such as the emission of ultra-violet from a display, as causing skin rashes. As Dr Tjønn had illustrated, there may well be other factors which were giving rise to these kinds of skin effects.

He pointed out that Dr Rycroft went into detail about ultra-violet but into no other measurements, perhaps highlighting the ultra-violet emission while not mentioning, in so much depth, some of the other measurements that were taken at the same time. Dr Rycroft agreed that the measurements he quoted were of ultra-violet because that was the wavelength that he was most familiar with, as a dermatologist, as causing skin rashes.

A national official of a trade union asked if the most likely cause of these skin rashes was a drying of the skin due to the extra heat emitted by the VDTs in an air-conditioned or air-heated atmosphere. VDTs emit a considerable amount of heat, and only in a very few offices have they been compensated for by adapting the environmental condition when the VDTs were introduced. Dr Rycroft said that he had certainly been involved in cases of red faces amongst office workers who had not been working with VDUs which had been alleviated by raising the humidity and lowering the temperature. This might explain some of the problems.

S. G. Allen of the National Radiological Protection Board, said that the measurements that they made at the factory were of radiation emissions not of static. F. Harlen, also of the National Radiological Protection Board said that they had taken measurements on other screens and found quite high static fields, as on television sets. They had visited the factory, before the Employment Medical Advisory Service, at the request of the company who made the VDUs and formed the same impressions as did EMAS concerning the workers' responsibility to their jobs. One operator had tried simple glazing for protection which he found ineffective, but double glazing was very good indeed. He also reported that he had erythema if he spent much time in his garden at the weekend. The skin condition of the other two men did not look particularly healthy, but it was not especially bad. The workplace was unpleasantly dry and slightly too warm when visited, and there was very little daylight coming through the windows, largely because of the screens used. The two operators who did not use double glazing used the 45-degree approach to their screens. One said that he found moisturizing creams very effective.

Tom Stewart, said that he had measured static fields from VDUs and found that the actual amount of charge radiated from a VDU was of the same order that one would find from a photocopier onto a piece of paper. In a normal environment the charge just dissipates. He suspected that the low relative humidity levels spoken of could mean that the people in that environment would build up a charge, and would therefore have dust, etc. attracted to their skin. If they were particularly sensitive, he believed the rashes mentioned could easily be induced. Dr Rycroft agreed that dusts are moisture-absorbing and that this might possibly cause dehydration of the facial skin which could subsequently lead to itching and erythema. A local government medical adviser asked if the operators concerned had any previous history of skin disorder or any history of respiratory disorders such as asthma, to which Dr Rycroft replied that they did not. In response to an engineer from a computer supplier it was stated that no allergy tests for dust, computer printout ink, clothing, paper, etc. had been carried out on these operators.

A director of medical services from a commercial company asked about the incidence of skin rashes in the company, and was told that there were only three or four men out of sixty who were complaining of these symptoms. He said that in his previous job he had had at least 1200 people using VDUs in air-conditioned offices. If there were two employees with skin rashes in a population of sixty he would have expected a very large incidence in a population of 1200, and there were no reports from this company of any skin rashes.

A health and safety officer from an insurance company asked Dr Tjønn whether any eye examinations were carried out on VDU operators in Norway. Dr Tjønn confirmed that they were concerned about eye problems. Some of the people who had developed skin problems also complained of red eyes but the redness of the eyes was slight and disappeared quite soon after leaving work.

In reply to another question Dr Tjønn said that where the rashes had appeared near the eyebrows or where there was a concentration of hair there had been no loss of hair. Asked whether it was possible that in some of these cases at least, the rash developed because the individual felt stressed or harassed in the work place Dr Tjønn replied that it might be an explanation but added that if a person is stressed in their work they might work more constantly and more intensely in front of the VDU screen. On the other hand, when one was stressed other biological changes also occur; with the hormones, blood vessels, and the skin's blood supply etc.

A health and safety officer from an insurance company asked Mr Cox (1) whether the HSE planned to make sample surveys at installations where VDUs had been in use for some years and which were perhaps not very carefully maintained; (2) what was the physiological significance of exposure to radiation below a frequency of 3 GHz; and (3) whether readings from the front, for the operator, and from the rear, which could effect a person other than the operator sitting near the equipment, are the most relevant positions for testing. Mr Cox replied that they had no plans to conduct any more surveys at present. The Scanning Spectral Radiometer which the NRPB were using no longer functions, so they were in some difficulties until the new instrument being put together was completed. Many of the other surveys carried out elsewhere had been done in work situations rather than in manufacturers' or suppliers' premises. The NRPB had performed private surveys on 'in-use' VDUs, and while that information was confidential between them and the customer, he thought that these surveys had not shown up any problems. Therefore he could not justify spending more money on this project, as it had already been very expensive. As regards the physiological effects of low frequency, Mr Cox replied that 3 GHz is a fairly long way up the scale if one is going to measure down to a few kilohertz. Certainly, one would be traversing regions where there is certainly the potential for a biological effect. Concerning the testing position, he chose those where he could obtain the maximum reading, which depends upon the circuitry of the box. This was not done on all of the devices as it was costing the best part of £1000 per VDU to have this ERA survey done. However, all the results were consistent with each other. Measurements were taken where the transformers were located in the case and from the operator's position. The line transformers are not necessarily located at the back or the front; in the ones tested they were in various positions under the casing.

In reply to further questions Mr Cox confirmed that they had looked at just about every device which was available at the time or soon to be marketed, which included colour sets as well as monochrome. He added that the BRH in the USA were arranging for a number of VDUs in a test house effectively to be blown up. They were going to 'wind up' everything until they just blew apart and hopefully measure all that was occurring, such as high voltage loadings etc. In the HSE survey the wick was wound up to obtain the brightest spectrum; but they did not go beyond that. It was felt that the operator would not be using the device if the picture were lost.

Dr Tjønn asked about the measured radiation emissions from incorrectly operating VDUs. Mr Cox replied that he was unable to make measurements on faulty VDUs. In a survey it was very difficult to obtain such VDUs, with the exception of the one that happened to break down on the ERA survey at Leatherhead. All the other sets were adjusted brighter than they normally would have been. He could not think of any fault condition which could generate very large amounts of radiation under circumstances where the operator would still continue to use the set. This was always the problem with a survey where there is no access to a workshop with broken VDUs. He could not afford new VDUs but had to use other people's. Mr Cox was asked whether the electromagnetic fields from VDUs could do any harm to people using pacemakers. Mr. Cox replied that there were a great many types of pacemaker, and they have a diversity of circuitry which are susceptible to different effects. He did not see any problems in terms of the field strengths found in the vicinity of VDUs. There may be problems at very much higher field strengths, but not at these levels.

Professor Grandjean asked whether the use of positive presentation of characters, that is, with dark characters, would change the characteristics of the emission? Mr Cox replied that it would obviously change the visible radiation emission but he did not see it changing anything else.

A local government safety officer asked how far the tests described by Mr Cox paralleled those required to be carried out by the manufacturers under Section 6 of the Health and Safety at Work Act. Mr Cox replied that the manufacturers have duties to do a variety of things under Section 6, and in most cases one would turn to the manufacturer and have these sort of data brought forward. However, in order to cover the field entirely it was felt, since it was such a sensitive issue, that the HSE would mount the survey in association with the NRPB, that kept the whole thing on a constant footing so the results could be interrelated. Although it was the manufacturers' responsibility, because the answers were wanted quickly, the HSE had to bear the cost and do it. He had had some results from manufacturers and they had not revealed differences in terms of higher numbers. Some of the numbers had been sketchy in parts of the spectrum and the manufacturers had used different measuring instruments or had farmed the job out. Section 6 does not say how the tests must be carried out.

A medical adviser from a public utility said that as Mr Cox was concerned with the measurements of various radiations that arise from visual display units, which are virtually identical to those from domestic colour television sets and also from television sets used in schools, perhaps the point made about controlling the time that people were allowed to work with VDUs in industry should also apply to people watching television at home and to children at school. Tom Stewart stated that to compare the viewing of television for watching programmes with using a text display for reading and understanding texts was to misunderstand the great importance of the task. Most people have their television sets at the far end of the room, and do not spend much time concentrating on details on the screen. The viewing conditions are entirely different and they are in no way comparable tasks. Television is reasonably well designed for viewing television programmes; it is not especially well designed for viewing text.

Brian Pearce wished to correct the assumption that he stipulated a 4-hour limit. He said that it was understandable that unions would seek a maximum of 4 hours on-screen time where there were doubts relating to specific direct and indirect health hazards. While there were these doubts, and while relevant research studies and the implementation of the known solutions were not undertaken in Britain, it was understandable that unions would continue to seek rest-pause criteria and maximum on-screen time limits until some attack on these problems was made.

Professor Grandjean felt that the reduction of exposure should not be used as an alibi for not making improvements. He feared that the unions, once they had achieved the 4-hour limit, would continue to use unsuitable work-stations.

A national official of a trade union said that she was a representative of a union which recommends that its members should not work longer than 4 hours on screen. At present, very few of her members worked on-screen, but that did not mean that her union was not thinking ahead and planning for this. They were not using the time limitation as an alibi or as a cover-up for a Luddite policy of opposing technology but were seeking to limit the time that their members work on VDU screens, which many experts, define as a visually demanding task and one which poses certain hazards and problems for eyesight as well as possibly facial dermatitis. Because her members' jobs depended upon their being able to see she did not want them exposed to risks until those risks could be eliminated or diminished. If they damaged their eyes they would have no job, and in the long run the unions must defend its members' interests. It was in their interest not to overwork on a screen until these problems were eliminated. She was pleased to hear appeals for more research and for the results of the research to be more widely circulated and discussed and kept in true perspective. But until that was the case she did not want her members working for eight hours at an unknown risk.

A safety officer from an insurance company asked how important it was to have a glare index of 16, which had first of all to be attained and then maintained. Tom Stewart replied that the measurement of the glare index in the UK was a peculiarly British phenomenon. Glare was measured differently in other countries. The numerical value was not of fundamental importance. The glare index attempts to quantify the degree of discomfort, irritation, and annoyance that people experience when getting unwanted light shining in their eyes. The important aim is to minimize this unwanted light. The glare index is usually calculated using artificial light, but glare from the sun can often be far worse. He had heard people say that a terminal must have a glare index of 16. This well illustrates how a little knowledge can be a dangerous thing, as the glare index refers to lighting and has nothing to do with terminals.

He reminded delegates that the transcript of the 1978 conference in Loughborough, far from being outdated by recent developments, was actually still entirely relevant and a useful source of information. (*An edited transcript of the one-day meeting on Eyestrain and VDUs* is available from The Ergonomics Society, c/o Department of Human Sciences, University of Technology, Loughborough, Leics LE11 3TU, £4.50.)

Dr Zaret said that when Prof Grandjean raised the subject of opacities and cataracts he failed to make the distinction between them: Dr Zarat emphasized that he spoke only about cataracts. There are several orders of magnitude of difference in their physical size but, more importantly, the physical location is very definite in the cataracts that he mentioned. Cataract is a clinical designation of an interference with visual function. Dr Zarat explained that he had undertaken the first, and probably the most extensive, study of cataloguing opacities in the lens. He compared a group of microwave workers with a control group and found that, lineally with age, there was an increasing number of minute opacities in the lens. He also discovered that these opacities had nothing to do with cataract formation.

Peter Stone, Reader in Vision and Lighting, Loughborough University asked Dr Zaret whether he had also observed radiant energy cataracts among people who had not worked with VDUs or among office workers. Dr Zaret confirmed that he had never seen this type of cataract in young people who had not been associated with equipment emitting these various radiations, but this did not mean that such cases did not exist. There could be cases where the cause is unknown because the person is unaware of it. Most of the cases he had seen had been referred to him by other eye specialists who had already made the diagnosis. Dr Zaret added that he also practises general ophthalmology and does not generally have people walking in from the street with these types of cataract.

When he started his career in ophthalmology thirty years ago, he hardly ever saw this type of cataract. The nature of cataract was changing and the incidence increasing. We were also seeing the age of onset diminishing. These findings were occurring in parallel with the ever-increasing atmosphere of electronic smog occasioned by the electronic revolution. That, in itself, did not prove the association. However, it would not be prudent to ignore these trends, especially as there were no other viable explanations. Milton Zaret explained that all his cases, including the man aged fifty-four, were too young to have naturally occurring cataracts. There is a statistical incidence that starts around the age of sixty-five to seventy of about 1% and this increases slightly with age. It never gets to 100%. However, when people in their fifties, forties, thirties, or even twenties have cataracts and work with radiant energy, and one had ruled out any other cause, it is assumed that the radiant energy had contributed to, if not caused, those cataracts. One has, of course, to rule out all other causes of cataracts; for example, uveitis, an inflammation inside the eye, can cause cataracts in some people. In the case studies cited, the cataracts are asymmetrical, for example, the two eyes are different, or the cataracts are not at the same stage of development. Any person with a mixed etiology has been excluded, the cases retained only differ in the various types of radiation to which they were exposed.

A safety officer from a computer supplier asked Dr Zaret about the setting of standards to which manufacturers have to abide when producing visual display units. His company was one of those producing to the UL standard of the United States. He wondered whether Dr Zaret had any influence on the UL.

Dr Zaret felt that adoption of the UL Standard was a mistake. He had been on two standard-setting commissions but had resigned from them because it was impossible to change the standards that had already been delineated. He sympathized with manufacturers but he was not on the committee which determined the American standard for microwave radiation. One distressing aspect was that there were two scientists on that committee whose own research showed that $3-4$ mW cm^{-2} would produce testicular injury, and yet they set the standard higher than that. He had been unable even to change the language, let alone the numbers in the standards. He did not wish to become involved in the technical details of what was wrong with some of the American standards but he had sympathy with those who are living by them. He would not rely on them for safety. Regardless of the standards laid down, if the eye has received an amount of radiation that is injurious, the eye will express the injury. The eye does not listen to what the standards say.

Dr Zarat was asked if ophthalmologists could say that they knew what had caused a cataract unless they knew what could have caused it. Dr Zarat agreed that if someone was presented to them with a cataract they would not be able to state whether it was a microwave cataract or some other type without performing a differential diagnosis. One could only diagnose a microwave cataract if one knew that the patient had been exposed to microwaves.

Further questions strongly disputed Dr Zarat's interpretation of the US standards for microwave, RF and lasers. In particular his interpretation of the permitted level for continuous exposure contained in the microwave standard was the subject of a heated exchange. Professor Grandjean concluded the discussion by suggesting that there was an hypothesis of some relationship but that twelve cases were no statistic and no proof.

Section 2

Section 7

Chapter 10

Issues in Vision and Lighting for Users of VDUs

P. T. STONE

*Reader in Vision and Lighting, Department of Human Sciences,
Loughborough University of Technology, Loughborough, UK*

The symposium on Eyestrain and VDUs (Stewart, 1978) served to highlight some of the problems related to the visual health and comfort of operators of VDUs. Since that symposium a considerable number of papers have been published as a result of research which has investigated many of the problems more fully. In this chapter I would like to review, in the light of this additional experience, what has been clarified and what practical recommendations might be possible in relation to the two issues of (A) eyestrain and vision testing and of (B) lighting requirements.

EYE STRAIN AND VISION TESTING

Incidence

One of the consequences of the growth of VDU usage has been a massive concern with complaints about eye strain, almost as though it were a recently discovered phenomenon. It has, in fact, been recognized for centuries and an observation by Plautus (251–184 BC) summarizes the condition: 'Sitting hurts your loins; staring, your eyes'. It is desirable, therefore, to keep the topic in perspective and to remember that it is a complaint that arises from many types of working conditions, and we must ask whether there is a higher than normal incidence of eye strain among users of VDUs.

A survey on office workers employed at traditional office tasks, conducted by Brundrett from the Electricity Council in 1974, showed that about 40% of the 600 staff studied complained of eye strain. Goacher, an ophthalmic optician, reported in 1980 to the National Lighting Conference that symptoms of eye strain were experienced by 50% of his patients attending for ophthalmic examination and for 25% of them eye strain was their sole complaint. They were not necessarily people who worked with VDUs. He concluded that evidence from various sources indicated that eye strain affects about a third of the working population.

In recent studies by different researchers comparisons have been made between VDU and non-VDU work tasks of the incidence of complaints of eye strain and related discomforts. Rey and Meyer (1980) found that 'VDU operators complained more often of visual impairment than other operators of the same age' and that the differences are statistically significant.

It was particularly noticeable that a high incidence of complaints arose from operators working 6–9 hours in front of the screen and fewer complaints arose from those working for periods of 4 hours or less. Läubli *et al.* (1980) show complaints of eye fatigue (and I am interpreting from their bar charts for this data) of 50% among traditional office workers, 60% for operators using data entry terminals and 70% for operators on conversational terminals. These data include complaints varying in incidence from several times a week to several times a month. Ghiringhelli (1980) interviewed 62 VDU operators and 237 females in different jobs not using VDUs. He found among other things, that what he referred to as 'eye diseases', (we must translate that to include discomfort, fatigue etc.) affected VDU operators significantly more (p<0.01) than other clerks. Cakir (1978), in the Ergonomic Society Symposium, reported that 80–90% of people doing work at VDUs complain of eye strain.

The facts from these more recent observations, which we did not have available in 1978, therefore strongly support the contention that working at a VDU does give rise to eye strain more frequently than do other kinds of work. The question that now arises is, what are the causes?

Causation

The terms eye strain or visual fatigue are used widely and often cover a variety of symptoms as may be seen from the various papers already quoted. I prefer to employ the term 'visual discomfort' since it makes no specific inference about symptoms or causation. Under this general term I would like to summarize the causes of visual discomfort, as an interaction between operator variables and environmental variables. The operator variables include ocular factors such as refractive error, convergence insufficiency, accommodation fatigue and binocular instablity.

In addition to the role of these refractive and muscle balance conditions in visual discomfort there are other sensitizing factors. Some people in the population are especially sensitive to discomfort glare. Dr Debney has found in her research at Aston University that many cases of migraine are actually visually induced. Also there are those people who have an intense dislike of light and many indeed suffer pain. Often this is a transient condition which may afflict a person with eye injury or corneal irritation. Hence it will be found occasionally that some VDU operators are likely to be much more sensitive than usual to lighting and display conditions.

The next part of the analysis on visual discomfort relates to the environmental factors, and clearly lighting is an important part of that analysis. Glare from light fittings is one source of discomfort, as are specular reflections from the task and flicker from the lamps. It is important to remember that the thermal environment may have an effect as humidity, high air temperature and air movement can all affect the cornea of the eye and give rise to discomfort.

It would be quite impossible to consider all of the relevant factors in detail so I intend to discuss in the following section those ocular and environmental factors that seem to be of key relevance to visual discomfort.

Ocular factors

Among the most frequent dysfunctions of the eye found in the population are the well-known refractive disorders of myopia, hyperopia, astigmatism, and presbyopia. Without adequate correction by means of spectacles these conditions could certainly create visual discomfort with prolonged visual work. However, eye strain sensation may still arise even when people have normal vision or perfectly adequate spectacles.

In his *System of Ophthalmology* Duke-Elder (1970) presents a chapter on eye strain and draws particular attention to insufficiency or excess of accommodation by the lens of the eye on the one hand and muscle balance and convergence difficulties of the extra-ocular muscles on the other.

Generally, these ocular processes work in close relationship with one another, although being capable of a degree of independence, and much of what we understand as eye strain can arise from an overtaxing of these systems attempting to produce a clear image in the brain. Thus the focusing processes in each eye together with the fusion of images from both eyes to form a single perceptual experience are essential functions for comfortable viewing. Duke-Elder makes the further important point that part of the fatigue experienced in eye strain may be due to a tiring of higher mental processes such as interest and attention when 'the call is made upon them to interpret blurred and indistinct images'.

These ideas have been appreciated for a long time but the difficulty has been to establish that these processes are in fact changed by near visual work and to find appropriate testing procedures to demonstrate such changes. It is my contention that both clinical observation and experimental work in recent years can make us much more confident that eye strain comes about as a direct consequence of the visual system trying unsuccessfully to maintain a clear visual image through accommodative effort and extra-ocular muscle activity and striving to gain fusion of the images from the two retinae. It is extremely important to note at this point that individuals complaining of eye strain vary in the way in which their own systems are stressed under visual work conditions and simple screening tests will not necessarily reveal the processes responsible, especially when the individual is not in a fatigued state.

Recent work by Ostberg (1980) using laser optometry where one can observe the actual activity of the lens, has shown that significant changes in the accommodative state of the lens occur after prolonged work on VDUs. His subjects, who had worked at real tasks, became after 2 hours more myopic (near-sighted) for relatively distant stimuli and more hyperopic (long-sighted) for relatively near stimuli. In other words, the actual accommodation required for the particular viewing distances used did not match the expected accommodation. Haider *et al.* (1980) found that after a 4-hour work period on VDUs, with breaks included, a temporary myopia was induced and it took 10–15 minutes to gain good distance vision after the work had finished.

Elias *et al.* (1980) noted a greater incidence of blurred vision and reduced acuity among a group of VDU operators who were using the equipment more intensively than a group with less usage. Dainoff (1980) shows a 24.9% increase of complaints of 'blurry vision' among twenty-three library cataloguers, after their work on VDUs where the average looking time was 75%. This group had a significant increase in post-work complaints of eye strain. It is of particular interest to note here that he was unable to find significant differences from visual screening tests before and after work. This serves to show the complexity of the system involved in eye strain and that standard tests do not necessarily reveal the functions that are involved.

The above comments are made in relation to problems of accommodation. The other function of central interest is that of muscle balance and binocular co-ordination.

Close visual work has been known to induce conditions of muscle imbalance which may result in either an esophoria or an exophoria after the working period. Sutcliffe (1950) drew attention to this and stressed the need to consider the accommodation-convergence relationship in eye strain, and not just accommodation alone. In a recent publication we at Loughborough University have made reference (Stone *et al.*, 1980) to the changes in muscle balance before and after work on a difficult visual task, where in effect the convergence system for the two eyes shows a tendency not to return to its pre-work posture. That finding may be closely linked to the subject of convergence insufficiency which has been commented on in relation to visual fatigue by Duke-Elder, and especially by Mahto (1972). This condition is a failure to maintain a comfortable convergence for near work, with one eye tending to move away from the true point of convergence. Thus what is being emphasized here is the importance of a satisfactory relationship between the eyes in terms of muscle control and lens activity.

One aspect of muscle balance has just been discussed but there are a number of different features of binocular unco-ordination that may give rise to visual discomfort. This, however, is a most complex topic that cannot be easily summarized here. I will just simply point out that there are people who have difficulty in maintaining a stable binocular alignment for near vision tasks, with the consequence that they cannot maintain image stability within the corresponding retinal areas that they habitually use to attain fusion. They may find that they adapt to a particular distance and that there is a distinct difference in visual comfort when viewing to one side of the midline as compared with looking to the other side. The process of attaining a comfortable fusion of the images from both eyes may be placed under stress if a new viewing posture is adopted, such as increased viewing distance.

Many copy typists state that they are more comfortable when their copy is on the left rather than on the right. That may seem to be a trivial need, but it arises from the fact that binocular co-ordination and stability are better in one viewing direction for some people. Poor co-ordination would give rise to a blurring of vision, loss of binocular acuity and stereopsis, and a consequent increase in visual discomfort.

These conditions can be exacerbated by a lack of cerebral dominance. Considerable professional experience is required for the assessment of these anomalies of binocular co-ordination and they may not be revealed by an ordinary muscle balance test or by vision screening devices. This subject has been well explored by Bedwell (1972a, b, 1978), who recommends (1972a) in addition to other possible methods of investigation for visual discomfort, the use of fixation disparity tests, the cover test, and dynamic retinoscopy. This does mean that professional expertise is required for such examinations which go well beyond the simple proceedures for ordinary visual screening practice. In this context it is appropriate to quote one ophthalmic optician: 'After attempting to use both mechanical screeners and a reduced clinical technique for screening, the author has concluded that, on the whole, a full visual examination is the safest and the most helpful way of ensuring that individuals are visually suited to the job and that VDUs do not harm the eyes' (Levitt, 1980).

Headache

The symptoms of eye strain may be accompanied by complaints of headache, well known as a referred condition. It is specifically illustrated by Läubli et al. (1980) in relation to work on VDUs. They derived two factors from a factor analysis of a self-rating questionnaire, applied to personnel using VDUs and to typists and office workers. One of these factors accounted for 81% of the variables, and they referred to it as eye fatigue or eye irritation. The minor factor, accounting for 19% of the variables, related to impaired accommodation. Within the first factor we can see a substantial weighting on headache (see Table 1).

Table 1. Factor analysis of eye impairments among 295 office workers (Läubli *et al.*, 1980)

		Loading
	Pains	0.71
	Burning	0.66
Factor 1	Fatigue	0.64
81% of	Shooting pain	0.53
variable	Red eyes	0.49
	Headaches	0.42
	Blurring of near sight	0.79
Factor 2	Flicker vision	0.62
19% of	Blurring of far sight	0.45
variable	Double images	0.45

Carlow (1976) classifies headaches arising from ophthalmic factors into three main categories: refractive disorders and muscle imbalance; disease of the eye and orbit; referred pain. In the classic treatise on headache by Wolff (1972) the observation is made that close application of the eyes as in reading, fine drawing, and close work in poor light may cause headaches in the frontal or occipital regions in the head and be accompanied by feelings of fatigue or strain in the eyes and by blurred or double vision. Both Carlow and Wolff attribute these symptoms to sustained muscular action with excessive accommodative effort and sustained contraction of the extra-ocular muscles in an attempt to produce single binocular vision with proper fusion. Head and neck muscle strain, due to an unsatisfactory head posture, can also be a significant cause of headache.

There are other causes of headache arising from afflictions of the eyes such as inflammation of the conjunctiva or iris, glaucoma, orbital tumour, optic neuritis, and herpes of the trigeminal nerve. Thus we should take note that in individual complaints, causes other than working conditions might be responsible for symptoms of headache.

In concluding this section attention is drawn to the following points:
(1) There is a considerable amount of evidence that visual discomfort is a common complaint and that its incidence seems to be higher than normal for personnel, especially those who work intensively and for prolonged periods at visual display units. There are many different causes of visual discomfort and they may arise from within the operator or from his equipment or from both.
(2) Complaints of eye strain are frequently associated with blurred vision and headache. The ocular causes underlying these complaints are not simple

one-factor processes but result from interactions within a complex dynamic system.

(3) Where the causes are due to accommodative and convergence strain, rest periods at the least, are clearly indicated, but in addition some operatives may need a spectacle prescription and even eye exercises.

(4) Where environmental and equipment factors are not the obvious causes of persistent complaint it is recommended that a professional examination is made by an ophthalmic optician and that the examiner should have knowledge of the operator's workplace and type of work.

(5) Vision screening tests are useful as a pre-work exposure record of the eyesight of personnel, and especially for detecting the common disorders of seeing. However, they are not adequate for diagnosing the complexities of anomalies in binocular co-ordination.

ENVIRONMENTAL FACTORS

In this second section I should like to draw attention particularly to the lighting and thermal factors that may cause visual discomfort.

Lighting Requirements

An important task consideration that has been gaining much attention in recent years is the role of veiling reflections, especially in printed tasks. It is well known that printed material on a glossy surface produces specular reflections which reduce the visibility of the print, and everyone is, of course, aware of the massive specular reflections from VDU screens that interfere with display contrast.

But even with a good copy which is not obviously glossy, discomfort can still be experienced because print, paper, or pencil frequently give rise to small specular reflections, especially under a non-uniform lighting condition. A measure which assesses the veiling reflection effect has derived mainly from American practice, and is known as the Contrast Rendering Factor (CRF). This indicates how well a lighting installation renders the visibility of a task in a specific position for a particular observing angle. A lack of visual clarity and discomfort is readily caused by a luminaire in the 'offending zone', that is, the area of a room surface which would be seen by the observer if his task were replaced by a mirror (Slater, 1979). This concept can be applied to the VDU situation to test whether some aspect of the discomfort might be due directly to luminaires coming into the 'offending zone'.

The point is that both VDU screen and copy need to be considered for visibility and comfort when lighting systems are being created. What is required now is a method for evaluating the CRF of VDU screens in relation to different lighting systems. The other source of discomfort from lighting systems is from the lamps and fittings themselves which often give rise to discomfort glare. Hence, there are two lighting design needs, the one to yield good visibility by reducing veiling reflections and the other to reduce lamp glare. In general, there are three broad ways of considering solutions to these problems.

Low-luminance Downward Directional Luminaires

This involves the use of low luminance fittings giving a confined, downward distribution of light, e.g. as in the louvred and optically controlled fittings, with a BZ1 or BZ2 type of distribution.

BZ values represent generalized angular downward distributions of light from luminaires, and the BZ1 and BZ2 classes provide narrow distributions. To avoid specular reflections on screen faces the VDU position needs to be related to the luminaire position and its light distribution.

Up-lighters

Indirect lighting has been known for many years but has rarely been employed for functional purposes. In offices up-lighters may be concealed in specially designed furniture using the ceiling to bounce the light around the room. In the headquarters of one large industrial concern in the UK up-lighters have been installed in their VDU offices with great effect. Fluorescent lamps and some of the new high-energy output lamps can be used in this system.

With almost any up-lighter system it is possible to get below the recommended glare index of 19 for general offices. When installing the system it is essential to consider the bounding surfaces which need to be highly reflective, but not glossy, to reflect the light effectively around the room.

Local Lighting

Desk lamps are sometimes proposed as an alternative or in addition to ceiling lights. I think it is a mistake to use local lighting systems, although some reasonably effective desk lamps have been designed for use with VDUs. Local light sources should not be required if offices are designed according to the standards laid down. Local lighting requires extra cabling and plug points, creates extra expense, and adds to the complexity of the situation and often generates glare for other occupants.

Combinations of the above may be possible, depending on the circumstances. A useful review of lighting for visual display units is now available from the Lighting Division of CIBS (1981).

Thermal Environment

In evaluating the environmental factors that cause visual discomfort at the workplace it is important to draw attention to thermal factors which may irritate certain surface tissues of the eye. The main causes are dehydration of the cornea due to high ambient temperatures and low humidity, conditions readily encountered in environments controlled for the computer (not for the occupants!) and sometimes a high air movement causes a draught across the cornea and dries mucous membranes. Basic environmental measurements on room temperature, humidity, and air movement need to be carried out when complaints arise about a 'stuffy atmosphere' and tired eyes. Cakir *et al.* (1980) report that in most offices the air is too dry rather than too moist and recommend that the relative humidity in office environments should be between 50% and 55%. They found that air movements of 0.1 m s^{-1} gave rise to an uncomfortable draught for 23% of people questioned and caused in some cases dryness and burning sensation in the eyes.

CONCLUSIONS

The points being emphasized in this chapter are as follows. Complaints of visual discomfort from working environments occur with significant frequency to be taken seriously and should not be dismissed as quirks of the imagination. The causes of such complaints are varied but they are amenable to investigation. They may be a characteristic of the operator, or of his environment, or an interaction between them. There are thus a number of physical assessments which can be made on the operator and on the environment to identify causes of complaint. These factors are summarized in Table 2.

Table 2. Summary of possible physical assessments for the evaluation of visual discomfort

Refractive state—conditions of long-sight, short-sight, etc.
Binocular co-ordination
Accommodation and convergence
Contrast Rendering Factor of tasks
Glare from lighting
Room temperature—dry bulb
Relative humidity
Air movement

It is also known that there are certain characteristics of displays that cause major discomfort problems, e.g. flicker, oscillating luminances of characters, contrast of display information with background, sharpness of characters, etc., but this topic is sufficiently extensive to merit separate treatment.

The remedy for these latter problems lies with the equipment manufacturers and in technological development. However, such development does require some guidance from human factors research in that standards need to be suggested, but in some areas information is still required. It seems that important questions here relate to the relative merits of positive and negative contrast displays; whether small bright screen characters give rise to disability glare—what I call veiling reflections from the task; and to determine at what frequency values images are perceived to be stable.

The essence of this chapter is contained in Figure 1. Various types of common input to the visual system, in terms of contrast, luminance, glare, etc., are represented on the left and the chief components of the responding system, which serves to optimize visual clarity are on the right. A lack of image clarity may occur due to certain characteristics of the input, or a defensive response against extremes of the input (e.g. glare) may be required. In these circumstances, the dynamics of the system will seek an optimum state. However, extreme situations may involve an extended posture for one or more of the elements and place the system under stress. This results in discomfort. The implication is that visual discomfort may often arise from multiple changes in the system rather than from a single variable. Thus vision testing needs to sample the various aspects of the system, especially when it is under load.

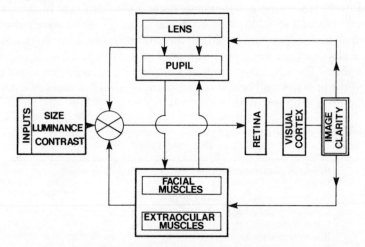

OCULAR SYSTEM RESPONSE DYNAMICS TO TASK INPUTS

Figure 1. A diagrammatic representation of the interaction between the operator and the environment

I have attempted in this chapter to bring into focus certain issues that a few years ago were rather diffuse and remained as questions. Although not all questions pertaining to eye strain have been answered, it does seem that we have some ideas about its causes and that working environments for VDUs can be improved using present knowledge about eyesight, equipment design and lighting.

REFERENCES

Bedwell, C.H. (1972a). 'Anomolies of the binocular system, accommodation and convergence, and discomfort for near vision', *Transactions of the International Ophthalmic Congress*, (1970) The British Optical Association, London, pp. 320–23.
Bedwell, C.H. (1972b). 'The eye, vision and visual discomfort', *Ltg Res. & Technol.*, **4**, (3), 151.
Bedwell, C.H. (1978). 'Binocular vision and near visual performance and fatigue', in *Current Concepts in Ergophthalmology*, ed. by Tengroth *et al.*, Societas Ergophthalmologica Internationalis, Stockholm.
Brundrett, G. (1974). 'Human sensitivity to flicker', *Ltg Res. & Technol.*, **6**, (4), 205.
Cakir, A. (1978). 'The incidence and importance of eyestrain among VDU operators', *Eyestrain & VDUs Conference*, The Ergonomics Society, Loughborough, UK.
Cakir, A., Hart, D., and Stewart, T.F.M. (1980). *Visual Display Terminals*, John Wiley, Chichester.
Carlow, T.J. (1976). 'Ophthalmic causes of headache', in *Pathogenesis and Treatment of Headache*, ed. by O. Appenzeller, Spectrum Publications, Jamaica.
CIBS (1981). *Lighting for Visual Display Units*, Technical Memo No. 6, Chartered Institution of Building Services, Delta House, 222 Balham High Road, London, SW12 9BS.
Dainoff, M.J. (1980). 'Visual fatigue in VDT operators', in *Ergonomics Aspects of Visual Display Terminals*, ed. by E. Grandjean and E. Vigliani, Taylor & Francis, London.
Duke-Elder, Sir Stewart (1970). 'System of ophthalmology', in *Ophthalmic Optics and Refraction*, Vol. V: Section IV, Eye strain and visual hygiene, Henry Kempton, London.
Elias, R., Cail, F., Tisserand, M., and Christmann, H. (1980). 'Investigations in operators working with CRT display terminals: relationships between task content and psychophysiological alterations', in *Ergonomic Aspects of Visual Display Terminals*, ed. by E. Grandjean, and E. Vigliani.
Ghiringhelli, L. (1980). 'Collection of subjective opinions on use of VDUs', in *Ergonomic Aspects of Visual Display Terminals*, ed. E. Grandjean, and E. Vigliani.
Goacher, J. H. (1980). 'Eyestrain: the environmental causes and their prevention', CIBS National Lighting Conference, March 1980, University of Kent, Canterbury, UK.
Haider, M., Kundi, M. and Weissenböck, M. (1980). 'Worker strain related to VDUs with differently coloured characters', in *Ergonomic Aspects of Visual Display Terminals*, ed. by E. Grandjean, and E. Vigliani.
Läubli, Th., Hünting, W., and Grandjean, E. (1980). 'Visual impairments in VDU operators related to environmental conditions', in *Ergonomic Aspects of Visual Display Terminals*, ed. E. Grandjean, and E. Vigliani.
Levitt, J. (1980). 'Visual problems associated with the introduction of visual display units into existing commercial offices', CIBS National Lighting Conference, March 1980, University of Kent, Canterbury, UK.
Mahto, R.S. (3 June 1972). 'Eye-strain and convergence insufficiency', *British Medical Journal*, **564**.
Östberg, O. (1980). 'Accommodation and visual fatigue in display work', in *Ergonomic Aspects of Visual Display Terminals*, ed. by E. Grandjean and E. Vigliani.

Plautus, T. Quoted from B. Ramazzini's *De Morbis Artificum*, from the translated edition by W.C. Wright (1964). *Diseases of Workers*, Hafner Publishing, New York.

Rey, P., and Meyer, J.J. (1980). 'Visual impairments and their objective correlates', in *Ergonomic Aspects of Visual Display Terminals*, ed. by E. Grandjean and E. Vigliani.

Slater, A.I. (1979). 'Variation and use of contrast rendering factor and equivalent sphere illuminance', *Ltg Res. & Technol.*, **11**, (3), 117.

Stewart, T.F.M. (ed.) (1978). 'An edited transcript of the one day meeting on Eyestrain and VDUs', The Ergonomics Society, Department of Human Sciences, Loughborough University, Loughborough, UK.

Stone, P.T., Clarke, A.M., and Slater, A.I. (1980). 'The effects of task contrast and visual fatigue at a constant illuminance', *Ltg Res. & Technol.*, **12**, (3), 144.

Sutcliffe, R.L. (1950). 'Visual work, visual comfort and visual efficiency in the weaving industry', *J. Physiological Optics*, **7**, 16–29.

Wolff, H. (1972). *Wolff's Headache and other Head Pain*, Revised by D.J. Dalessio, Oxford University Press, New York.

Chapter 11

More Practical Experiences in Solving VDU Ergonomics Problems

TOM STEWART

Butler Cox & Partners, London, UK*

The development of the silicon chip as a powerful, flexible, and cheap systems component has encouraged the convergence of the technologies of computing, telecommunications, and office equipment. Increasingly, this convergence results in new products being developed which are more complex and sophisticated than their pre-convergence ancestors. For example, word processors offer vastly superior facilities for the creation and processing of text than the electric typewriter. Stored program controlled PABXs provide the user with powerful facilities for manipulating telephone calls which extend the effectiveness of telephone communications.

Yet even with these impressive developments the human problems frequently remain unsolved. Some staff refuse to operate VDU-based word processors for fear of damaging their eyes. Many desks are unsuitable for use with computer equipment and cause users to adopt awkward and constrained postures. Frequently SPC PABX users cannot use the powerful facilities because the cumbersome and tedious procedures for accessing them are impossible to remember.

The considerable potential of the technology is not being realized because the human interface is unsatisfactory, and it is often difficult to make equipment usable or understandable. These are not new problems. Machines have been poorly designed for human use for many years both in the factory and in the office.

However, VDU ergonomics has attracted considerable publicity (mainly adverse) in recent years. This publicity is often out of all proportion to the problem, or out of proportion to other problems which are ignored. I have

* Now with Systems Concepts Limited, London, UK.

walked through factories full of noise, dust, and noxious fumes, past unsecured ladders and dangerous equipment to reach an office in a corner. Here it is relatively quiet, and has a pleasant thermal environment where workers sit at comfortable desks using keyboards.

I have been called into these offices because they are worried about the health hazards of VDUs. The fear and uncertainty which surrounds computing has amplified the problems that do exist until they are out of all proportion. This has put pressure on users, trades unions, managers, and computer suppliers to find out the facts and to act on them. Unfortunately, the research results are often difficult to interpret and may even appear contradictory. The researchers themselves are often unwilling or unable to commit themselves to specific recommendations. Some manufacturers offer advice but they are not unbiased; they are probably not going to tell you lies but they might not tell you the whole truth.

Many organizations therefore need independent, realistic, and practical evaluation of their VDU problems followed by specific advice and recommendations. Management consultancies offer advice which is based on evidence, experience and findings from investigations and can be of invaluable help here.

The purpose of this chapter is to review our experiences in tackling such VDU evaluation assignments for a number of clients during 1979 and 1980. Only VDU evaluation assignments for user organizations are reported, although in each case far more than just the VDU hardware was evaluated. Typically, the workplace, environment, working methods, and organization were also assessed.

APPROACH AND METHODS

Butler Cox & Partners is a totally independent management consultancy concerned with computers, telecommunications, and office automation. My own background includes nine years of research and teaching ergonomics at Loughborough University before joining the company as a specialist in the human aspects of information technology. The methods used for the investigations are based on research conducted throughout Europe and reported more fully elsewhere. (For example, see Cakir et al. (1980).) The methods included:

(1) Conducting detailed interviews with the relevant staff;
(2) Observing the working procedures and working environment;
(3) Measuring directly the work-station dimensions, illumination levels, as appropriate;
(4) Distributing and analysing brief questionnaires;
(5) Analysing existing company records and documentation.

The checklist in Cakir et al. (1980) was used as a basis for much of the measurement and assesment. However, the criteria were not regarded as rigid, and considerable interpretation was necessary to establish the significance of the findings. As a consequence, the present analysis only deals with the problems which were quite clearly significant for the user. We look for major exceptions and deviations and relate those to the tasks of the operators. This combination of equipment assessment and task analysis is of fundamental importance, as recommendations cannot be simply translated to other situations and tasks.

SAMPLE OF VDU ERGONOMICS CLIENTS

The sample used for this review contains all our user clients who commissioned VDU ergonomics evaluations during 1979 and 1980. How representative it is cannot readily be established. In each case the organization at least had identified that it needed to prevent or reduce VDU problems. This could mean that we only experience organizations where there are problems and that elsewhere everything is perfect. However, I believe that since our clients have taken the trouble to call in outside help, this is a clear indication of their awareness of the problems. They are therefore more likely to have considered ergonomics when installing their own VDUs. The problems elsewhere are likely therefore to be at least as significant. The sample composition is shown in Table 1 and the types of application in Table 2.

Table 1. Survey sample

20	Applications
7	Organizations
450	Users (approximately)
320	VDU workplaces

Table 2. Applications and user types surveyed

Centralized data entry	4
Word processor operators	1
Full time clerical operators	4
Intermittent 'professional' users	5
Tele ordering operators	2
Machine operators	4
	20

The role of trades unions in the assignments is shown in Table 3. Unfortunately trades unions are not always involved with these assessments, although we do recommend that managements discuss the results with them as early as possible.

From our experience it is clear that many problems are nipped in the bud when trades unions are involved right at the beginning.

When the equipment or the environment does not meet the needs of the operators, or the task they have to perform, the results often show themselves as complaints of headaches, eye strain, etc. But they may also lead to excessive errors, reduced productivity, delays, inefficiency, and waste. Workers, as well as management, are concerned with these problems and it is not unusual for operators to complain that technology prevents them from doing their jobs as well as they would like. In the current economic climate productivity and efficiency matter to all.

Table 3. Trade union involvement

	Reporting to Union		
Initiative for assignment	Joint	Separate	None
Union	2	1	—
Management	2	—	2

TYPICAL VDU FINDINGS

The two most important VDU features assessed were the quality of the displayed image and the characteristics of the keyboard. These are considered in turn below.

Image Quality

The quality of the displayed image on a CRT based VDU is a complex function of several parameters. Table 4(a) lists some of the most important character formation parameters and shows the checklist answers for the fifteen models evaluated. In some cases, the VDU did not conform to the value shown in the checklist but was nonetheless found to be acceptable in the practical situation. These items are shown as 'OK'.

Inadequate spacing between adjacent characters and between successive rows was the most frequently encountered problem (67%). In many cases, the system designer could only produce acceptable formats by introducing blank lines between rows, greatly and excessively reducing the capacity of the display.

The shape of the characters was not acceptable in five cases. These VDUs ignored typographic recommendations on the thickness of lines, the width/height ratio and the formation of individual characters.

Poor resolution of the characters was a common complaint. This was not caused by faulty design of the display but usually by dust being drawn in and deposited on the back of the screen, making the characters illegible. Contrast problems can be caused by worn cathode ray tubes. This reduces the contrast between the characters and the background.

Table 4(b) lists some of the luminance and stability requirements and shows the checklist answers. Character instability includes flicker, swim, and any other apparent movement in the displayed image. Despite the claims of virtually every manufacturer to produce 'flicker free' displays, distracting instability was a problem on six of the fifteen models. From the tables it can be seen that virtually all the terminals had a refresh value of 50 Hz and yet still did not produce a stable image. Half of the models had some problems with the luminance and the resolution of the characters. Typically, on an inadequate display, one had to be traded off against the other. Individual operators varied in their response to this compromise and sacrificed whichever seemed less important to them. Where there was insufficient adjustability in the position of the display the users consequently had to adjust their own posture, giving rise to the physical problems that one would expect.

Keyboard

The keyboards were assessed on a number of parameters and the results are summarized in Table 5. Table 5(a) is concerned with general keyboard criteria, Table 5(b) with the characteristics of individual keys and Table 5(c) with overall keyboard layout. The tables show that the majority of keyboard characteristics were acceptable. The most frequent shortcomings concerned the keyboard thickness and its detachability. Both of these were observed to be causing constrained and awkward postures in the users. One often has to choose between reading the display at a comfortable distance or keying comfortably and efficiently. In some situations, the users had attempted to lower the keyboard by cutting a hole in the desk. However, this made it impossible to move and reduced the leg room underneath the workplace.

The layout of the keyboards was assessed in relation to the specific task of the user. In many cases the layout of function keys could be improved but the existing layout could not be considered a major problem. One exception to this occurred in a data entry application where there were eleven variations of layout and function among fourteen apparently identical keyboards. Such attempts to confuse the operators seem almost deliberate. Mirror-like reflections were also found from some of the key tops.

TYPICAL WORKPLACE FINDINGS

A total of twenty-four different designs of workplace were evaluated and the results are summarized in Table 6. I use the term 'designs' of workplaces rather generously: I wish they had been designed, as in many cases they were simply

Table 4(a). Typical display findings for 15 models of VDU

Character formation

	1	2	3	4	5	6	7	8	9	10	11	12	13	14	15
1. What is the colour of the characters in the display?	green	blue/white	white	green	green	green	white	green	green	white	white	white	black on white	green	white
2. Is the character height greater than or equal to 3 mm?	Yes not fixed	Yes 5 mm	Yes 3 mm	Yes 3 mm	Yes 3 mm	Yes 3 mm	Yes 6 mm	Yes 3 mm	Yes 3.5 mm	Yes 4.5 mm	Yes 5.5 mm	Yes 5 mm	No 2 mm	Yes 4 mm	Yes 4 mm
3. Do the individual dots merge to produce a sharp image?	No	Just	Yes	Yes	Yes	Yes	Yes	Yes	Yes	OK	Yes	No	Yes	Yes	Yes
4. Dot matrix	5X7	5X7	5X7	7X9	7X9	5X7	5X7	5X7	7X9	5X7	7X9	5X7	7X9	7X9	7X9
5. Is the character width 70–80% of the upper case character height?	not fixed	OK 60%	Yes 80%	Yes 72%	No	OK 66%	OK 66%	OK 66%	Yes 70%	No 56%	No 40%	OK 60%	Yes 75%	OK 60%	OK 62%
6. Is the space between the characters between 20% and 50% of the characters height?	not fixed	Yes 40%		Yes 40%	Yes 40%	OK 16%	OK 16%	OK 16%	Yes 22%	No 16%	No 10%	Yes 30%	No 10%	Yes 20%	OK 15%
7. Is the row spacing between 100% and 150% of the character height?	not fixed	Yes 100%	OK	Yes	No 80%	No 60%	No 42%	No 60%	Yes 114%	No	No	No 60%	No 75%	Yes 100%	No 50%
8. Is it possible to adjust the screen about its horizontal axis? (screen angle) *on desk	No	No	No	No	No	No	No	No	Yes*	No	Yes	No	Yes	No	Yes
9. Is the upper edge of the screen at or below eye height	Yes	Yes	OK	No 450 mm	No 480 mm	Yes 300 mm	Yes 360 mm	Yes 250 mm	Yes 400 mm	Yes 310 mm	Yes 405 mm	Yes 250 mm	OK 420 mm	Yes 400 mm	Yes

Table 4(b). Typical display findings for 15 models of VDU

Display screen luminance

	1	2	3	4	5	6	7	8	9	10	11	12	13	14	15
1. Is the character luminance greater than 45 cd m^{-2} and less than 160 cd m^{-2}?	Yes	Yes	Yes	Yes	Yes	OK	OK	OK	OK	OK	OK	No	OK Rev.-video	Yes	OK
2. Do the character images remain sharply defined at maximum character luminance?	No	Yes	N/A	Yes	No	No	OK	Yes	Yes	No	No	No	OK	Yes	Yes
3. Is the background luminance adjustable?	Yes	No	N/A	No	Yes	No	No	No	No	No	Yes	Yes	Yes	Yes	No
4. Is the contrast between the characters and background acceptable?	OK	OK	OK	OK	OK	OK	OK	Yes	Yes	OK	OK	No	OK	OK	Yes
5. Are the displayed character image stable?	No	OK	OK	Yes	Yes	No	No	No	Yes	No	No	Yes	Yes	Yes	Yes
6. Is the refresh rate at least 50Hz with negative display (light characters, dark background) with low–medium persistence phosphors?	Yes	Yes	Yes	Yes	Yes	Yes	Yes	Yes	Yes	Yes	Yes	OK	80Hz Rev.	OK	Yes
7. What is the surface of the screen	Etch	Etch	Plain	Plain	Etch	Etch	Etch	Etch	Thin film	Fine etch	Mesh	Plastic	Plain	Plain	Etch

Table 5(a). Typical keyboard findings for 15 models of VDU

General keyboard criteria

	1	2	3	4	5	6	7	8	9	10	11	12	13	14	15
1. Is the keyboard detached from the display screen console, i.e. joined by a cable?	No	Yes	No	Yes	Yes	Yes	No	No	Yes	No	Yes	Yes	Yes	Yes	Yes
2. Is the weight of the keyboard sufficient to ensure stability against unintentional movement?	—	Yes	—	Yes	Yes	Yes	—	—	Yes	—	Yes	Yes	Yes	Yes	Yes
3. Is the thickness of the keyboard, i.e. base to the home row of keys, less than 50 mm? (acceptable, 30 mm preferred)	Yes	Yes	No	No 70 mm	No 78 mm	No 72 mm	Yes 30 mm	No 70 mm	No 75 mm	No 77 mm	OK 60 mm	No 68 mm	No 75 mm	No 85 mm	Yes 50 mm
4. Is the angle of the keyboard in the range 5–15°?	Yes	Yes	Yes	Yes 11°	Yes 11°	Yes 10°	Yes 9°	Yes 11°	Yes 10°	Yes 7°	Yes 13°	Yes 10°	Yes 11°	Yes	Yes 7°
5. Is the surface of the keyboard surround matt finish?	Yes	Yes	N/A	Yes	Yes	OK	OK	OK	Yes	Yes	Yes	OK	Yes	Yes	Yes
6. Is there at least a 50 mm deep space provided for resting the palms of the hands?	Yes	No	Yes	Yes 55 mm	Yes	No 15 mm	OK 45 mm	No 15 mm	Yes 60 mm	No	No	Yes 50 mm	OK 30 mm	OK	No

Table 5(b). Typical keyboard findings for 15 models of VDU

Key characteristics

Key characteristics	1	2	3	4	5	6	7	8	9	10	11	12	13	14	15
1. Is the key pressure between 0.25 and 1.5 N?	Yes	Yes	Yes	Yes	Yes	Yes	Yes	Yes	Yes	Yes	Yes	Yes	Yes	Yes	OK
2. Is the key travel between 0.8 and 4.8 mm	Yes	Yes	Yes	OK 5	OK 5	Yes 3	Yes 4.8	Yes 4.8	Yes 3	Yes 4.0	Yes 3.8	Yes 4	Yes 4.8	Yes	Yes 4.5
3. For square keytops is the keytop size between 12 and 15 mm?	Yes	Yes	Yes	Yes 14	Yes 14	Yes 12	Yes 13	Yes 13.5	Yes 12.5	Yes 12.5	Yes 12.5	Yes 12.5	Yes 12.5	Yes 12.5	Yes
4. Is the centre spacing between adjacent keys between 18 and 20 mm?	Yes	Yes	Yes	Yes	Yes	Yes 19	Yes 19	Yes 20	Yes 18.5	Yes 19.0	Yes 19.0	Yes 19.5	Yes 19	Yes 18.5	Yes
5. Are the keytop surfaces such that specular reflections are kept to a minimum?	No	No	N/A	Yes	No	Yes	Yes	Yes	Yes	Yes	Yes	No	No	Yes	Yes
6. Is the activation of each key accompanied by a feedback signal:															
Audible click?	No	Yes	N/A	Yes	No	Yes	Yes	No	Yes	No	No	Yes	No	Yes	Yes
Tactile click or snap action?	No	No	N/A	No	No	No	No	No	No	No	No	No	No	No	No
7. Is the keyboard provided with a roll-over facility:															
2-key roll-over?	Yes	No	No	No	No	Yes	No	No	No	Yes	Yes	No	No	No	No
n-key roll-over?	No	Yes	No	n-key	n-key	No	Yes	No	Yes	No	No	Yes	Yes	Yes	Yes

Table 5(c). Typical findings for 15 models of VDU

Keyboard layout

	1	2	3	4	5	6	7	8	9	10	11	12	13	14	15
1. Does the layout of the alpha-keys correspond to the conventional typewriter keyboard layout?	Yes	Yes	Yes	Yes	Yes	Yes	Yes	Yes	Yes	Yes	Yes	Yes	Yes	Yes	Yes
2. Does the layout of the numeric-keys—above the alpha-keys—correspond to the conventional typewriter keyboard layout?	No	No	Yes	Yes	Yes	Yes	Yes	Yes	No	Yes	Yes	No	Yes	Yes	Yes
3. Are the numeric-keys grouped in a separate block?	No	No	Yes	Yes	Yes	Yes	Yes	No	Yes	No	No	No	No	No	Yes
4. Are the most important function-keys colour-coded?				No	Yes	No	Yes	No	No	No	No	No	Yes	No	No

Table 6(a). Typical workplace findings

Checklist item	Answer	
1. Are sufficient number of work surfaces provided?	YES NO	83% 17%
2. Are the working surfaces of sufficient size?	YES NO	92% 8%
3. Are all items of equipment and job aids which must often be manually manipulated within the normal arm reach of the operator?	YES NO	62% 38%
4. Is the height of the keyboard above floor level between 720 and 750 mm?	YES NO (worst case 825 mm)	25% 75%
5. Is the height of the leg area sufficient?	YES NO	91% 9%
6. Is the leg area at least 700 mm deep?	YES NO (worst case 330 mm)	75% 25%
7. Is it possible for the operator to easily re-arrange the workplace, e.g. by changing the positions of the VDT and other items of equipment?	YES NO	44% 56%
8. Is a footrest provided?	YES NO (not a problem for all users)	8% 92%
9. Are the electrical supply cables and other services to the VDU and workplace adequately secured and concealed?	YES NO	65% 35%

accidents. The workplaces developed according to where the sockets were, where the equipment had been initially placed, or where a corner could be found. They were not tailored to the operators who had to use them or designed to provide space for documentation.

The most frequent problem was lack of adjustability in the work-station. The majority of the chairs had some adjustability but it was usually difficult to use. Typically it involved getting off the chair and exerting considerable force on a knurled knob. As a consequence, these chairs were seldom adjusted. Only a few workplaces had a gas-lift chair action, although these were much preferred and exploited by the users. People will adjust their workplaces if they are shown how the adjustments can be made, but many users were unaware of the adaptations possible. Few of the workplaces were designed for VDU use and hence in many cases the keyboards were too high (up to 825 mm in some cases) when placed on an ordinary desk.

Cable management systems need to be more flexible so that they can be used realistically. It is sometimes necessary to move equipment, but this usually results in cables lying along the floor. This is a potential danger not only to people but also to the equipment as well.

Leg-room was insufficient in many workplaces due to obstructions under the desk or to the desk-top being too thick. It was reduced to a depth of 330 mm in the worst case. Insufficient leg room was not only found in adapted workplaces but also in equipment especially designed for VDUs.

Table 6(b). Typical seating findings for 24 different types of workplaces

Checklist item	Answer	
1. Is the seating height easily adjustable?	YES	71%
	NO	29%
2. Is the height of the backrest adjustable?	YES	75%
	NO	25%
3. Can adjustments be made easily and safely from the seated position?	YES	9%
	NO	91%
4. Is there guidance available to the individual operators to help them achieve an optimum adjustment to their chair?	YES	—
	NO	100%
	(not a problem for every operator)	

Many of the workplace problems were aggravated by the design of the VDUs, as mentioned above. Thick fixed keyboards made the shortcomings of the workplaces more critical. Heavy fixed screens also reduced the ability of the user to adapt the workplace to their own requirements.

TYPICAL ENVIRONMENT FINDINGS

The environment of all but one of the different types of workplace were assessed and typical findings are shown in Table 7. In almost all cases some aspect of the thermal environment was reported as causing problems. In some locations direct measurements were made but in the rest, the staff and management evaluations were noted. The introduction of VDUs and ancillary equipment added considerably to the heating and ventilating problems in some organizations. However, not all the complaints can be attributed to this. The sedentary nature of VDU work may make some of the users more aware of deficiencies in the environment. The environment may also be a convenient scapegoat for other complaints related to the VDU or to computers in general. Nonetheless, the heating and ventilation systems evaluated were seldom capable of controlling the temperature, airflow, and humidity to an adequate degree even when operating perfectly (which was seldom the case).

Workplaces were often draughty and there were often significant temperature differences between the air at head level and floor level. This was partly caused when inadequate systems were overloaded when the new equipment was installed, resulting in massive amounts of hot or cold air being moved about in order to balance the temperature. Heating and ventilation are difficult to optimize, because of individual preferences, and the thermal environment often becomes a major source of complaint. In all the locations there had been problems in this area, although they were not necessarily related to the VDU.

Noise from the VDUs was only a problem in a few cases (typically from the cooling fan), although some younger operators reported being distracted by the high-pitched whine from the electronics. Printer noise was frequently disruptive.

Table 7. Typical environment findings (for 34 different environments)

Checklist item	Answer	
1. Is the work room air conditioned?	YES	24%
	NO	76%
2. Have steps been taken to avoid local hot spots, e.g. under desks, in corners, etc?	YES	44%
	NO	56%
3. Is the noise level less than 65 dB(A) in routine task areas?	YES	85%
	NO	15%
4. Is the noise environment free from high frequency tones?	YES	82%
	NO	18%
5. Are there other items of equipment in the workroom e.g. printers, teletypes which generate high or distracting levels of noise?	YES	50%
	NO	50%
6. Is the illuminance between 300 and 500 lux?	YES	53%
	NO	47%
	(minimum found 54 lux, maximum 2500 lux)	
7. Are there glare sources in the operator's field of vision, e.g. lights, windows etc?	YES	66%
	NO	34%
8. Are the luminaires equipped with prismatic or grid type glare shields?	YES	60%
	NO	40%
9. Are the windows fitted with internal blinds?	YES	65%
	NO	35%

It was not typically at a high enough level for major complaints but it often interfered with communication or concentration.

A frequently encountered problem concerned reflections and glare from windows and luminaires. The illumination level itself was less of a problem. Just over half of the workplaces had an illumination level on the desk surface of between 300 and 500 lux (which is suitable for screen and document reading).

Most of the problems concerned excessive illumination (up to 2500 lux in some cases), although the opposite extreme also occurred (54 lux). Glare from unshielded or poorly shielded light fittings and from windows, was a major cause of problems in the visual environment.

OTHER FINDINGS

In the organizations studied there were a number of other problems which were often as important as the more traditional ergonomics issues. These included personnel and supervision problems, difficulties in monitoring and controlling workflow and unfriendly or awkward man–machine dialogues. Similarly there were organizations where potential problems had never materialized because these other issues had been handled successfully. For example, where VDU work was only a part of a well designed job, problems with the hardware were seldom significant.

Sometimes if the job has the right psychological and social ingredients people will often not complain about ergonomic problems and will tolerate bad equipment. Computer programmers will often subject themselves to the most incredible conditions when motivated to solve a particular problem. Of course, lack of complaints does not necessarily mean that all is well, that they are working most effectively or efficiently. They would probably be even more productive and satisfied if these defects were removed. The hard economic reality is that we need to get more efficiency out of many computer operations. This can be solved cheaply by improving the man–machine interface, allowing the equipment to work for the users rather than against them.

RECOMMENDATIONS FOR SOLVING THESE PROBLEMS

The purpose of the above studies was to solve or prevent problems. A total of twenty sets of recommendations were made and the most common types of recommendations are summarized in Table 8.

Table 8. Typical recommendations

Recommendations (N = 20 reports)	
Improve workplaces	75%
Replace chairs	65%
Modify lighting	70%
Make minor VDU change	40%
Make major VDU change	15%
Reposition VDUs	40%
Conduct operators' eye tests	15%
Modify ventilation	20%
Modify organization	15%
Reduce printer noise	10%

The recommendations had to be practical and realistic. Regrettably, in some cases this meant that there was little that could be done about some of the problems. Problems can be ameliorated sometimes by altering something entirely different. For example you can rearrange the pattern of work by allowing full-time data entry workers to move about to collect their work, so constrained postures are no longer a problem. However, modification of the work organization is something that is often difficult. There were many occasions when we would ideally have liked to recommend that the VDUs were scrapped for a particular reason but could not economically justify that action even to ourselves. Nonetheless there were three occasions when the evidence justified major action of this kind. More frequently we recommended improving the workplace and modifying the lighting.

Many of the modifications were relatively simple, involving additional work-surfaces or a different position of existing desks or luminaires. Sometimes specially designed adjustable VDU work-stations were recommended, however, only where there was little or no adjustability in the equipment itself. Few of the fully adjustable workplaces on the European market are entirely satisfactory. Most involve trade-offs between ease of use and stability (especially when loaded with heavy VDUs).

The most practical way to introduce some adjustability into the workplace is often through the chair. Well designed, easily adjustable, stable chairs are widely available. The adjustability is more likely to be used than the nominal adjustability in expensive VDU desks.

Modifications to the lighting involved both changing luminaires and repositioning, although in some office designs this is impractical. Simply removing excess fluorescent tubes results in uneven light distribution but can be an acceptable compromise. One effective way of reducing illumination on source documents is to provide a document holder. Holding the documents at an angle also improves the keyboard/screen/document relationship. However, over-elaborate holders tended not to be used and we found simple lecterns most effective.

Natural lighting was made more suitable by recommending blinds or curtains, although this was not always satisfactory. When the visual environment could not be made satisfactory various filter attachments were recommended. The various types available all have their disadvantages (in terms of reduced resolution, luminance, or high cost).

Modifying the ventilation usually meant providing some additional control over ineffective automatic systems. Most of the solutions were far from perfect since many of the defects were basic design faults and would have involved major changes.

Where there was still doubt about the ergonomics of the workplace or where the planned use was particularly intense, we recommended operator eye tests (VET Advisory Group, 1980). The cost of screening or tests is relatively minor while the reassurance value is very high for many people. There are many cases where they are entirely justified although it is misleading to say that every VDU operator needs an eye test. One organization tested the eyes of all its sixty operators, having previously agreed to buy any special spectacles that were found necessary. Four operators needed prescriptions and because of their initial responsible attitude the price of those spectacles has brought industrial peace to the company.

CONCLUSIONS

As I pointed out earlier, it is difficult to decide whether this sample is at all representative of other VDU users. My own view is that the problems we find are all too common to many VDU users, although clearly there are some who are better off and some who are worse. Nonetheless, there are several general conclusions which we believe can be drawn from our experience.

(1) There are many problems which can be solved with existing ergonomics knowledge—provided that it is applied. The major shortfall is in application rather than derivation of ergonomics knowledge.

(2) VDU manufacturers are beginning to improve the ergonomics of their products. The recommendations are gradually being applied and many new products offer good ergonomic features such as thin detachable keyboards and clear easily adjusted screens.

(3) Workplace and environment design lags behind and is the cause of many VDU problems. Far too many workplaces are cramped, awkward, and inefficient, and too many environments are not conducive to effective, comfortable work. Adjustable workplaces which are on the market are often either excessively complicated, difficult to adjust, or unstable and therefore dangerous. More imaginative use of lighting is still sorely needed.

(4) Applying ergonomics need not be very expensive or very difficult but it helps if it is applied early. Applying ergonomics late involves modifications and unnecessary expense which makes justification difficult. It is often no dearer to buy a good chair than a bad one.

(5) Joint VDU ergonomics initiatives by management and trades unions are to be welcomed. Frequently such co-operation prevents subsequent recrimination and suspicion and ensures better systems for everyone.

(6) Ergonomics advice must be practical and realistic if it is to be taken seriously—but no one should use the weaknesses in current ergonomic knowledge as an excuse for inaction.

(7) The final conclusion is that over the eleven years I have worked in the area of VDUs and ergonomics the one thing I can say with absolute certainty is that if you try to ignore the problems, they do not go away, they increase.

Acknowledgements

Some of the work reported here was undertaken by my colleague Andrea Caws.

REFERENCES

Cakir, A., Hart, D.J., and Stewart, T.F.M, (1980). *Visual Display Terminals*, John Wiley, Chichester.
VET Advisory Group (1980). *Eye Tests and VDU Operators*, Final Report of the VET Advisory Group 1980, London.

Acknowledgement

REFERENCES

Chapter 12

Trades Unions and Ergonomic Problems

BRIAN PEARCE

The HUSAT Research Group, Loughborough University of Technology, Loughborough, UK

Over the last five years much has been published concerning the ergonomics of using VDUs. However, in Britain it is the trades unions rather than the manufacturers, purchasers or government bodies who have taken the initiative in promoting the application of ergonomic principles to the design and use of VDUs. This is not to deny that a number of other organizations have contributed to the increased awareness of the need to consider ergonomic issues.

I believe that the application of existing ergonomic knowledge can avoid or remove the great majority of injurious or deleterious effects that can arise from working at a VDU. I undertake consultancy work in industry and it is obvious from my own experiences that very little of our existing ergonomics knowledge is implemented in many of the computer systems in use at the moment. It is very important that this knowledge is disseminated and put into practice so that these mistakes and problems can be avoided.

A small number of computer manufacturers have ostentatiously promoted the use of ergonomic recommendations not only in their brochures but also in their equipment, e.g. DATASAAB: *Ergonomics the Third Factor.* The Business Equipment Trade Association (BETA) has published *A Guide to Users of Business Equipment, Incorporating Visual Display Units.* No doubt many large organizations using VDUs have developed their own in-house recommendations and guidelines for the selection and installation of VDUs. However, they are rarely made public.

At last government-related bodies in Britain have reacted. December 1980 saw the publication of the HSE *Research Paper No. 10,* 'Human factors aspects of visual display operation', (Mackay, 1980) and the HSE are in the process of producing a Guidance Note, the draft of which is currently being circulated.

It is also appropriate to mention the courses, conferences, and publications of the HUSAT Research Group. Members of the Research Group contributed to

The VDT Manual (Cakir *et al.*, 1980), originally published by IFRA, and a recent publication from the NCC: *Designing Systems for People* (Damodaran *et al.*, 1980).

However, by far the most prolific source of recommendations has been the trades unions. Notable among the early documents from the unions was that produced by ASTMS: *Guide to Health Hazards of Visual Display Units—An ASTMS policy document*. Other unions such as APEX, NALGO, AUEW (TASS), and the NUJ have produced documents on new technology which include ergonomic recommendations. The parts of the documents containing ergonomic recommendations vary in size, status, and purpose, from checklists to policy documents. Some of these confine themselves to what might be termed the health and safety issues while others consider the ergonomic issues as merely part of a broader policy on new technology. Because these documents have received such wide circulation I believe that they have had a profound effect upon increasing the awareness of many members of management, unions, and manufacturers of the need to consider ergonomic issues.

Inevitably, some of them incorporate union policy related to the prevention of job loss and de-skilling. Unfortunately, given the adversary traditions of industrial relations in Britain some of the so-called ergonomic recommendations are perceived to be politically motivated. From discussions with trade union members and officials it is apparent that some (but by no means all) of the health and safety recommendations are indeed influenced by what might be termed political factors. Unfortunately, one of the consequences is that extremely complex technical issues are translated into oversimplified recommendations, which then become the subject of negotiations between mangement and unions. Some of the negotiated settlements that have resulted merely demonstrate the ignorance of all the parties involved, e.g. the 57 Hz refresh rate negotiated in one new technology agreement when the union sought 60 Hz and management 50 Hz.

However, it is inevitable that some of the health and safety recommendations contain political implications. The introduction of new technology into the office is merely one part of what has been termed 'the technology revolution'. Inevitably, comparisons are often made between the technology revolution and the industrial revolution. One of the differences, however, is that unions as we know them today were not present during the industrial revolution. It is clear that trade unions will seek ways to avoid the consequences of the industrial revolution being repeated in the coming office revolution.

Typically there is a growing awareness of the need to avoid the creation of the full time VDU operator. There are many examples of such an operator in the history of data processing. Many systems have been designed with no thought whatsoever for the job content of the person who has to sit in front of the VDU all day.

Some unions, notably APEX, are very concerned about the content of the jobs of individuals working with VDUs. New technologies no longer require the centralized computing and data preparation facilities that typified data processing installations. Thus many of the technical constraints have been removed. However, to eliminate the concept of the full time VDU operator requires more than simply changing technology; it requires a fundamental change of attitude by those responsible for purchasing, designing and implementing the new computer systems and wide-ranging changes in organizational practices.

It is interesting to examine in more detail the various union approaches by considering a brief analysis of the coverage of the union policy documents. It is also interesting to note the changes that have occurred in the content of these documents.

APEX	*Office Technology—The Trade Union Response*. March 1979. *'Interim guidance for APEX negotiators'*.
ASTMS	*Guide to Health Hazards of Visual Display Units*. 1979. *'An ASTMS policy document'*.
AUEW (TASS)	*Health Hazards of Visual Display Units. 'Guidance to AUEW (TASS) Committee of Representatives'*.
BIFU	*Guidelines for the Operation of Visual Display Units, July 1978. '... Research departments Suggested guidelines on the use of visual display units'*.
GMWU	*Checklist for the Installation and Use of Visual Display Units, July 1979 (under review). 'Health and Safety Aids'*.
NALGO	*Visual Display Units (VDUs), May 1979. 'A NALGO checklist for negotiators'*.
SOGAT	*Interim Guidelines on Operation of Video Display Units within the Printing and Newspaper Industry*.

Figure 1. Union documents analysed (March 1980)

The framework within which the analyses have been conducted are based upon six headings: Hardware, Software, Physical Environment, Job Design, Training, and Planning of Technological Change. These headings refer to factors which have been identified by members of the HUSAT Research Group from over ten years' study of man–computer interaction as representing the major issues that need to be considered when examining the ergonomics of VDUs.

It cannot be emphasized too strongly that a consideration of the human aspects of VDU usage must include not only the traditional ergonomic issues such as features of the hardware, the physical environment, and the workplace but also the design of jobs, the ergonomics of the software interface, the provision of training, and the planning and implementation of technological change. It is all too easy to fall into the trap of prescribing that which is relatively easy to measure but ignoring that which is less easy to quantify.

It should be noted that there is no separate heading for the VDU user. This is intentional, since it should be fundamental to the ergonomist's approach to keep the person, the user, at the centre of their thoughts. Thus each of the six headings should be considered in relation to the user and the user interface.

There is no separate heading for ophthalmic or postural issues. This reflects the view that the 'symptom centred' approach has been responsible for the neglect of certain important issues concerning the ergonomics of VDUs.

Figure 1 shows the titles of the seven union documents that were included in the analysis in March 1980. The content of the documents were analysed using five categories (Figure 2). The more ticks under a particular heading against each document, the more that document includes on that topic. No ticks under a particular heading indicates that there is no reference to that topic in the · document. However, the categories are not intended as a qualitative judgement of the content but simply an indication whether the document presents the information with explanation or discussion or simply as a series of one sentence recommendations.

Detailed discussion	√ √ √ √
Explanation	√ √ √
Brief recommendation	√ √
Comment	√
No reference	

Figure 2. Categories used in the content analysis

Even though a document may contain detailed discussion of a topic it does not follow that it has covered all that an ergonomist might consider should be covered or that the information is completely accurate; issues that will be considered in more detail later.

It is immediately apparent from Figure 3 that the union documents included in the analysis in March 1980 gave very little attention to the ergonomics of the software. The physical environment received much the same attention as the hardware, and these two headings were the most covered by most of the unions. This no doubt reflects the pressures that caused the documents to be produced in the first place, that is, in response to the problems associated with visual and postural fatigue and the possibility of radiation hazards.

Six months later a similar analysis (Figure 4) presented a slightly different picture. The changes were brought about by the publication of new documents by TASS, NALGO, APEX, and NUJ (Figure 5). These new documents increased the coverage of the topics by TASS and NALGO, but it is interesting to note that the new APEX document *Automation and the Office Worker* actually reduced the emphasis on the ergonomic recommendations related to the hardware.

Discussions with officials from several unions suggested that they were aware

	Hardware	Software	Physical environment	Job design	Training	Planning of change
APEX	✓✓✓✓	✓	✓✓✓✓	✓✓✓✓	✓✓✓✓	✓✓✓✓
ASTMS	✓✓✓✓	✓	✓✓✓✓	✓✓✓	✓✓✓	✓✓✓
AUEW (TASS)	✓		✓✓	✓✓	✓	
BIFU	✓✓✓	✓✓	✓✓✓	✓		✓
GMWU	✓✓✓		✓✓✓	✓✓		✓
NALGO	✓✓		✓✓	✓		
SOGAT	✓✓✓	✓	✓✓✓	✓✓		

Figure 3. Content Analysis (March 1980)

	Hardware	Software	Physical environment	Job design	Training	Planning of change
APEX	✓✓✓	✓	✓✓✓✓	✓✓✓✓	✓✓✓✓	✓✓✓✓
ASTMS	✓✓✓✓	✓	✓✓✓✓	✓✓✓	✓✓✓	✓✓✓
TASS	✓✓✓		✓✓✓	✓✓✓	✓✓✓	✓✓✓
BIFU	✓✓✓	✓✓	✓✓✓	✓		✓
GMWU	✓✓✓		✓✓✓	✓✓		✓
NALGO	✓✓✓		✓✓✓	✓✓✓	✓✓	✓✓✓
SOGAT	✓✓✓	✓	✓✓✓	✓✓		
NUJ	✓✓✓	✓✓✓	✓✓✓✓	✓✓✓	✓✓	✓✓✓

Figure 4. Content Analysis (August 1980)

that there were some gaps in the coverage of the documents, but that this was due in some cases to a lack of resources and in others to an intentional emphasis upon issues over which they thought they could most easily have influence. These discussions also suggested that some of the unions were particularly concerned about the issue of job stress and the influence that the design of the software interface, the user's dialogue with the computer, might have upon the type and

APEX	*Automation and the Office Worker*, 1980
ASTMS	*Guide to Health Hazards of Visual Display Units—1979 An ASTMS Policy Document*
TASS	*New Technology—A Guide for Negotiators*
BIFU	*Guidelines for the Operation of Visual Display Units, 1978*
GMWU	*Checklist for the Installation and Use of Visual Display Units* (under review), '*Health and Safety Aids*', 1979
NALGO	*New Technology—A Guide for NALGO Negotiators, 1980*
SOGAT	*Interim Guidelines on Operation of Video Display Units within the Printing and Newspaper Industry*
NUJ	*Journalists and New Technology, 1980*

Figure 5. Union documents analysed (August 1980)

quality of the the job that the user was expected to perform. It was also very apparent during these discussions that some of the unions were very concerned that the issues that they needed to convey to their members were in some cases extremely complex and difficult to present. There was also an awareness that the numeric values quoted in the recommendations for certain features of the hardware and the physical environment were, in effect, diverting attention from other less quantifiable but no less important issues.

The two most recent union documents analysed are interesting in that they adopt very different approaches. In January 1981 there appeared via the NUM a document entitled *Guidelines for the Use of Visual Display Terminals in Word Processing Installations in the National Coal Board*. Unlike the other union documents mentioned so far which lay down policy, these NUM/NCB guidelines have been agreed between the National Coal Board and the clerical unions. In effect these guidelines are similar in status to the health and safety section of a new technology agreement. An analysis of the coverage of these guidelines shows that they concentrate solely on issues related to the hardware and the physical environment. There is no mention whatsoever of software or job design. Thus in comparison with other unions the guidelines cover far less than even some of the early union documents. It might be thought that the gaps in the coverage of these guidelines could be attributed simply to the fact that the unions concerned had little experience of using VDUs or lacked access to relevant technical expertise concerning VDU ergonomics. However, they cover not only the clerical members of the NUM but also members of other unions in particular APEX, which represents some clerical employees of the NCB. The reasons why the APEX guidelines have not been incorporated into the NUM/NCB guidelines are unclear.

The most recent document analysed, published by APEX (in February 1981), contrasts very strongly with the NUM/NCB guidelines. Entitled *New Technology—Health and Safety Guidelines*, the document is 'intended as a preliminary guide for health and safety representatives to help them assess possible health hazards of new technology and for staff representatives in negotiating about new systems'. In comparison with earlier documents from both APEX and other unions, there is a much greater emphasis on the software issues and job stress. The other notable feature is that, unlike earlier documents, it discusses the issues and objectives without recourse to a list of numeric values or technical specification for features of the hardware and the physical environment. This is a very interesting development that no doubt reflects the growing concern for the problems of interpretation that the traditional recommendations have caused for the members and officials of the unions.

As has already been mentioned, the almost obsessive concern for numeric values in many union recommendations has resulted in some bizarre negotiated solutions. Another problem is that, as technology changes, the technical specification of the ergonomic recommendations may need to be changed. If the recommendations are stated in terms of numeric levels rather than a human performance objective they may well preclude or delay the introduction of equipment that is ergonomically superior.

When selecting a VDU one of the most important ergonomic criteria should be that the screen has a 'clear, stable image'. There are many factors that influence the clarity, readability, legibility, and stability of a screen image and some of these factors are interrelated in very complex ways. Many of the union documents cite numeric values for a handful of the many factors that are known to influence the clarity and stability of the screen technology in common use. For example, the stroke/height ratio and the screen luminance levels that are commonly recommended for the positive contrast, negative image displays currently used are inappropriate for the negative contrast, positive image displays that are now being introduced.

Even if the appropriate numeric values are stated there is often a great deal of difficulty in assessing if a particular display meets the recommended criteria. This is a problem not only for the members of unions who seek to influence the selection of equipment but also for management purchasing the equipment. Many of the ergonomic recommendations relate to features that require specialist equipment and expertise to measure. Because of these difficulties requests are often made to the manufacturing companies to supply the information that for some reason they seem reluctant to put into their sales brochures. In many cases the manufacturers just do not know the answers.

An illustration of this problem is provided by a request to the HUSAT Research Group for advice from an earnest management services officer of a county council. He was trying to ascertain which of the four manufacturers of word processing equipment that had been short-listed for purchase complied with the recommended screen luminance levels. He had, in fact, drawn up a large table that included two sets of ergonomic recommendations for the equipment. One set of recommendations had been produced by the council's own health and safety officer and another by the NALGO representative. Incidentally the NALGO document *New Technology—a Guide for NALGO Negotiators* does not quote any specific screen luminance levels. The figure had probably been borrowed from other sources; a common situation.

The management services officer had contacted the four manufacturers and requested, among other things, the screen luminance levels. Figure 6 shows the replies he had received.

Jacquard	40 Lamberts
AES Wordplex	0–4 Foot-candles
Wang	7.3 Candela m^{-2}
Data logic	No reply

Figure 6. Screen luminance levels reported by manufacturers

Without going into too much technical detail it should be noted that Foot-candles are units of illuminance (i.e. not a measure of screen luminance) while the two figures that are quoted in units of luminance are somewhat improbable. An immense amount of time and effort must have been wasted by union officials, potential purchasers, and manufacturers in the pursuit of this ridiculous numbers game.

SOME SOLUTIONS

In reality many of the problems that have been described are due to a lack of understanding of the issues by unions, manufacturers, and management. There is a desperate need for all concerned to become more enlightened about the need for and implementation of ergonomic recommendations concerning VDUs. In the short term, however, there are a number of actions that would greatly reduce the problems.

(1) The unions who have issued ergonomic recommendations concerning VDUs should review their documents and follow the example set by

APEX to produce a simple, clear statement of the issues, problems, and solutions that will not tie their own members and officials in numeric knots. Apart from the new APEX document there is evidence that this process is occurring in other unions. It is also interesting to note that the white collar unions are discussing the publication of a joint document containing ergonomic recommendations concerning the design of VDUs. The HUSAT Research Group is also in the process of planning a series of tape/slide presentations on computer ergonomics that should help to explain the issues to both management and unions.

(2) Although the long term aim must be to eliminate the excessive emphasis on ergonomic recommendations that contain inadequate, ephemeral technical specifications, many of the existing problems would be solved if manufacturers made more information readily available. Manufacturers of equipment that utilize ergonomic recommendation should include the information in their brochures. They should specify the heat output, screen luminance levels, force/distance characteristics of the keys, and all the other factors that influence the ergonomic aspects of VDU usage. Those that do will sell more machines.

Ergonomic considerations are becoming an increasingly important item in the selection and purchasing decision of many users. Not to place this type of information in the sales brochures merely presents the potential purchaser with one more set of unknowns and one more reason for choosing an alternative supplier in a very competitive marketplace.

In the longer term there are a number of issues that need to be tackled.

(1) At present the main motivation of the unions for the implementation of ergonomic recommendations is the avoidance of postural fatigue, visual fatigue, and job stress. Some go further and seek improvements in the quality of working life for their members following the introduction of new technology. Laudable though these aims might be, they are not always shared by many employers. What is needed is proof, by means of a comprehensive cost benefit analysis, that the application of ergonomic principles not only removes the common problems but actually results in reduced operating costs which would amply offset any increased implementation costs.

It has long been the view of members of the HUSAT Research Group that there are hidden costs in failing to consider ergonomics in the development of a computer system. These may take the form of under-use or misuse of systems, high labour turnover, high error rates, increased training times, and reduced service to the organization's customers. Examples of all these hidden costs have come to light during the course of the research and consultancy work undertaken by the group. Unfortunately so far they have remained unquantified.

(2) The final suggestion for the reduction of some of the problems that have been mentioned concerns the establishment of a standards procedure or type approval scheme for VDUs.

The selection of ergonomically sound equipment is only one of the many ergonomic issues that should be considered when implementing a computer system. The selection of ergonomically sound equipment will not guarantee that operators will avoid postural or visual fatigue. However to select equipment that is not ergonomic will almost certainly guarantee that the users will suffer to some extent.

The ergonomic recommendations for VDUs, promoted in virtually all the documents published so far, rely upon the specification of certain parameters that are known to be correlated with the legibility of a VDU screen. However, it has already been mentioned that many of these recommendations fail to take account of the complex interrelationships between display characteristics and that changes in technology can require the rewriting of the specification.

One solution, therefore, would appear to be to dispense with specifying parameters that are known to correlate with legibility and adopt a direct measure or human performance-based criteria. Thus manufacturers would submit their equipment to a standardized test procedure utilizing a specified number of subjects. The performance of these subjects on a series of carefully selected tasks would indicate whether the display provided a suitable image. So long as the display reached certain minimum criteria it would receive a type approval. Such a procedure is described in a paper by Gorrell (1980). This would obviously have more credence if it were conducted by an independent testing authority. However, given the normal timescales needed by standards authorities in Britain such a procedure would be unlikely to be established for several years.

However, a human performance based standards procedure could be established more quickly if it received the general support of manufacturers, purchasers, and unions. Given the problems that have been described using the existing ergonomic recommendations, it would appear that a performance-based standards procedure could be of benefit to all parties. If there appears to be sufficient support from manufacturers, purchasers, and unions the HUSAT Research Group would consider establishing such a procedure.

BIBLIOGRAPHY

Union documents containing ergonomic recommendations

APEX (1979). *Office Technology—The Trade Union Response*, 'Interim guidance for APEX negotiators'.

APEX (1980). *Automation and the Office Worker.*

APEX (1981). *New Technology—Health and Safety Guidelines.*

ASTMS (1979). *Guide to Health Hazards of Visual Display Units—An ASTMS policy document.*

AUEW (TASS). *Health Hazards of Visual Display Units.*

AUEW (TASS). *New Technology—A Guide for Negotiators.*

BIFU (1978). *Guidelines for the Operation of Visual Display Units.*

GMWU (1979). *Checklist for the Installation and Use of Visual Display Units* (under review), 'Health and safety aids'.

NALGO (1979). *Visual Display Units (VDUs)*, 'A NALGO checklist for negotiators'.

NALGO (1979). 'Health and safety news'. *Health and Safety Information Sheet No. 6.*

NALGO (1980). *New Technology—A Guide for NALGO negotiators.*

NUJ (1980). *Journalists and New Technology.*

NUM/NCB (1981). *Guidelines for the Use of Visual Display Terminals in Word Processing Installations in the National Coal Board.*

SOGAT, *Interim Guidelines on Operation of Video Display Units within the Printing and Newspaper Industry.*

Other documents containing ergonomic recommendations

BETA, *A Guide to Users of Business Equipment incorporating Visual Display Units*, Business Equipment Trade Association.

Cakir *et al.* (1980). *Visual Display Terminals*, John Wiley, Chichester.

Damodaran *et al.* (1980). *Designing Systems for People*, NCC, Manchester.

DATASAAB, *Ergonomics the Third Factor.*

Gorrell, E.L. (1980). *A Human Engineering Specification for Legibility of Alphanumeric Symbology on Video Monitor Displays* (revised), Defence and Civil Institute of Environmental Medicine, Downsview, Ontario, Canada.

Mackay, C.J. (1980). 'Human factors aspects of visual display unit operation', *Research Paper No. 10. Health & Safety Executive*, HMSO, London.

Chapter 13

Postural Loads at VDT Work-stations

WILHELM HÜNTING

*Joint investigation with Dr T. Läubli and Professor E. Grandjean
Department of Hygiene and Applied Physiology, Swiss Federal Institute of
Technology, Zürich, Switzerland*

INTRODUCTION

Technological development and the application of electronic computer systems raises different problems. It is therefore important to recognize in good time the effects of technical innovations on man in order to prevent inhuman consequences. According to several authors, various complaints are attributed to the introduction of VDTs in the office. From the point of view of ergonomics three main aspects are recognizable:

(1) The psychological problems originate in the changed occupational activities. The integration of an office employee in a man–computer system influences the variety of his job and limits his social contacts. The work dictated by software restricts personal scope and acquires a repetitive character with a special demand on vigilance. As social contact between the workers is therefore reduced the VDT changes professional activities and the social position of office employees.

(2) Visual impairments are much higher in VDT operators than in traditional office work, as shown by our own field studies. They manifest themselves as functional problems which have to be attributed to the special eye strain due to VDTs. These symptoms may persist for more than a few hours, however, they are reversible.

(3) Constrained postures and overwork load in the cervicobrachial region and upper limb are the result of an integration of office employees in a man–computer system. These are seen in

(a) A forced fixed position of the head (view directed to screen and source document);

119

(b) A constrained posture of arms, hands, fingers and trunk due to a body posture which has to be adapted to machine or furniture;

(c) A fixed shoulder area due to repetitive finger movements (keyboard operation);

(d) A higher muscle tension through visual strain, mental load, and psychological adaptation.

Initially, long-lasting muscular tensions and repetitive movements may produce acute fatigue and pain in the muscles concerned (localized fatigue). These troubles are short-lived and reversible. If the static load is repeated daily over a long period of time, more or less permanent aches will appear; this second stage may also involve chronic fatigue such as physical impairments of tendons and joints. Symptoms of this kind were reported for keyboard operators and for VDT operators. It is reasonable to assume that inadequate workplace design might generate or aggravate such physical complaints.

This state of affairs was the background of a field study, carried out mainly in banks.

INVESTIGATED GROUPS

The study comprised two groups of VDT workplaces, one group of full-time typists and one control group of traditional office workers. Some characteristics of the four groups are represented in Table 1.

Table 1. The studied occupational groups (s = standard deviation)

	n	age $\bar{x} \pm s$	women	operators working $\leqslant$ 6h at keyboard or terminal
Data-entry terminal	53	30 ± 8	94%	81%
Conversational terminal	109	34 ± 12	50%	73%
Typing work	78	34 ± 13	95%	65%
Traditional office work	55	28 ± 11	60%	30%

Data-entry terminals

This work can be characterized as follows:

(1) Full-time numeric data-entry with the right hand;

(2) Work speed was high at about 12000–17000 strokes hr^{-1};

(3) The gaze was directed mainly at source documents;

(4) The documents were lying on the table.

Conversational terminals

Employees of this group carried out payment transactions in two different banks:
(1) Both hands operated a keyboard;
(2) The speed was moderate;
(3) The head was turned towards the terminal during about half of the time and towards the documents half of the time;
(4) The documents were lying on the table;
(5) In one bank, keyboard and terminal were movable on the table (conversational terminal A). In the other bank, keyboard and terminal were sunk into the table (conversational terminal B).

The conversational terminal A workplace contained a sufficiently large space to rest hands and arms, whereas at conversational terminal B only a narrow rim of about 3 cm to rest the balls of the thumbs was available.

Typewriter workplaces

In this group employees were occupied full-time with typing work, partly copying documents and partly using a dictating machine:
(1) Both hands operated a keyboard;
(2) Work speed was high;
(3) The gaze was mainly directed to documents lying on the left, or a dictating machine was used.

Traditional office workplaces

The task in this group was identical with the work at conversational terminals, i.e. payment transactions in a branch office of a bank which did not yet contain VDTs. The work can be described as follows:
(1) Keyboards were used only occasionally;
(2) There was great diversity of movements at the workplace.

METHODS

The following investigations were made:
(1) Inquiries on postural pains;
(2) Inquiries on work satisfaction;
(3) Measurement of workplace dimensions;
(4) Assessment of body posture at work;
(5) Medical examination of upper extremities, shoulders and neck;
(6) Taking forehead, hand, and room temperatures.

Inquiries on pains of the locomotor system were carried out by means of a questionnaire developed by the Japanese 'Committee on cervicobrachial syndrome of JAIH' (1973) and used in a random sampling study with 16.9 million people. The questionnaire was translated and modified and illustrated with anatomical drawings of trunk, hands and arms; it was checked by Maeda *et al.* (1980) on its suitability.

In the questionnaire, employees were asked for different parts of the body regarding
   (1) Pains;
   (2) Stiffness;
   (3) Fatigue (tiredness);
   (4) Cramps;
   (5) Numbness;
   (6) Tremor.

The answers were: 'daily', 'occasionally' or 'seldom or never'. Furthermore, the subjects had to judge their workplace and their work at the terminal; individual items were formulated as statements and five answer categories were given.

To investigate work satisfaction we used an abbreviated form of the questionnaire elaborated by Martin *et al.* (1980). The measured workplace dimensions are represented in Table 2. Body posture at work was measured with a method described by Hünting *et al.* (1980). The medical examination included an anamnesis and palpitation of painful pressure points in shoulders, arms and hands. Furthermore, painful symptoms in isometric contractions of the forearm were assessed. Skin temperatures of the forehead and right middle finger were measured some time during work.

RESULTS

Pains in the locomotor system

In Figure 1 the percentage of daily pains in the upper extremities as well as in back and lumbar region is shown.

The frequent occurrence of acute pains in arms, neck, and shoulders in the data-entry terminal group is of interest. The traditional office work control group is practically free of such symptoms. However, some especially striking results are:
   (1) Sensations of stiffness are observed, particularly in neck and shoulders;
   (2) Fatigue is localized in the right arm, and somewhat less frequently in the right hand;
   (3) Cramps and perception disorders occur in hands and arms, whereby the right side is more frequently affected.

All the recorded pains occur most frequently in the data-entry terminal group and practically never in the traditional office work group; the conversational terminal and typewriters groups occupy an intermediate position.

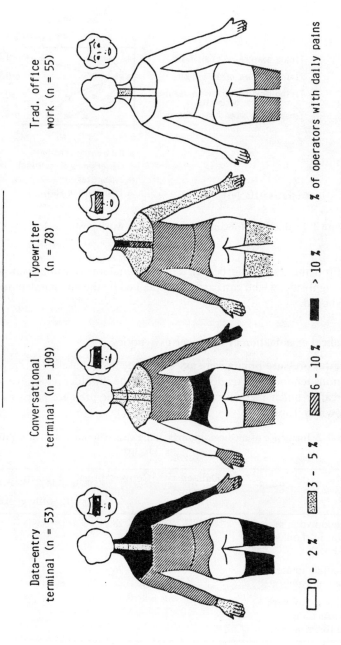

Figure 1. Incidence of daily bodily and eye pain in four different office jobs (100% = *n* of each group)

In the questionnaire the employees were also requested to judge the inconvenience of some body postures. The results are compiled in Figure 2.

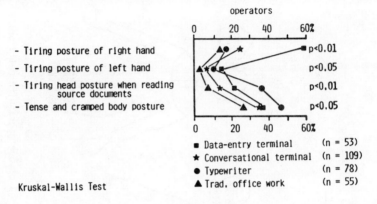

Figure 2. Evaluation of inconvenience of some work postures (100% = n of each group)

From the graph it can be seen that with data-entry terminals fatigue prevails on the right side, which can easily be explained by the one-handed operation of the keyboard.

Medical examination of the upper extremities

Painful pressure points were assesed in the neck muscles (m. trapezius), and in various tendons and tendon insertions in the shoulder region, which are classed hereafter under the general term 'tendomyotic pressure pains'. The results are shown in Table 2.

Table 2. Incidence of medical findings in the neck–shoulder–arm area (100% = n of each group)

	Data-entry terminal $n = 53$	Conversational terminal $n = 109$	Typewriter $n = 78$	Traditional office work $n = 54$
Tendomyotic pressure pains in shoulders and neck	38%	28%	35%	11%
Painfully limited head movability	30%	26%	37%	10%
Pains in isometric contractions of the forearm	32%	15%	23%	6%

About one third of the employees constantly working at keyboards show clinical findings in the neck–shoulder–arm region; in traditional office work such findings are much less frequent. Besides pressure pains, about half of the examined employees showed a strong tension of the neck muscles (myogelosis).

Table 3 shows the frequency of medical treatment of arm and hand pains during the past months and years.

Table 3. Frequency of employees having seen their doctor in the past few months or years about impairments in arms and hands

	n	%
Data-entry terminal	53	19
Conversational terminal	109	21
Trad. office work	55	13
Typing	78	27

The groups with frequent pains in arms and hands also consulted the doctor more often.

In the data-entry terminals group the employees with lower hand temperatures also showed a significantly increased frequency of pains in neck and shoulders and a greater occurrence of sensitivity disorders in hands and arms. Similiar findings were assessed by Nakaseko (1975) and Maeda (1977), who consider these symptoms as consequences of a compression of arteries and nerve tracks in the neck–shoulder region (cervicobrachial syndrome). These findings support the assumption that constrained postures at office workplaces may produce persistent injury involving inflammation and degeneration in the overloaded tissues. Jobs associated with high-speed keyboard operation seemed to enhance these effects.

We observed in our field studies inadequate designs of the workplaces as well as physical complaints. This led us to analyse the relationship between the work-station design and the physical complaints.

Workplace dimensions

In Table 4 the most important workplace dimensions for various groups are listed. The results show large table and keyboard heights. In all groups the office employees compensated for the high keyboards partly by raising the seat level, thus obtaining better distances between seat level and keyboard of between 28 and 33 cm.

Table 4. Dimensions of work-stations. Median values and 90 percentile (in cm). No working level was adjustable. (* Desks with sunk keyboards)

	Table height above floor		Keyboard height above floor		Keyboard height above table		Seat height		Keyboard height above seat	
	Median	90%	Median	90%	Median	90%	Median	90%	Median	90%
Data-entry terminal ($n = 53$)	71	67–78	78	75–87	7.8	7–9	48	45–55	30	26–36
Conversat. terminal A ($n = 55$)	76	70–78	84	73–87	8.6	2–9	51	47–56	33	25–37
Conversat. terminal B ($n = 54$)	76*		78*		2*		49	46–53	28	25–32
Typing ($n = 78$)	71	68–76	83	79–91	11.6	9–15	50	45–54	32	27–38
Trad. office work ($n = 55$)	70	69–75	81	75–83	9.8	5–13	49	45–53	31	24–36

Observed body postures at the workplace

Resting arms and hands

An essential characteristic of body posture is the support of arms and hands, which we have assessed by means of simple observations at workplaces with office machines. The frequency of resting is shown in Figure 3.

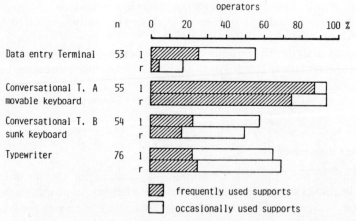

Figure 3. Incidence of supporting arms and hands (100% = n of each group)

It is striking that at the conversational terminals A with movable keyboard, hands and arms are frequently rested on the table, whereas this characteristic was much less frequent with the other groups. With the conversational terminal A the keyboard was, as a rule, placed in a way to leave a considerable space for support. With conversational terminal B as well as with the other two groups, the rest areas were rather narrow. This proves that arms and hands are frequently rested if adequate support is available.

Visual distances

In Table 5 mean values and the 90% ranges of visual distances to terminal and documents are listed.

Table 5. Mean values and 90 percentile of visual distances to the screen, to the typed text or to the source documents (in cm). ($\bar{x}$ = mean values)

	visual distances (in cm)			
	to terminal or written text		to documents	
	$\bar{x}$	90%	$\bar{x}$	90%
Data-entry terminal ($n = 53$) (movable keyboard and screen)	58	42–81	43	39–57
Conversational terminal A ($n = 55$) (movable keyboard and screen)	62	44–82	47	38–58
Conversational terminal B ($n = 54$) (fixed or built-in keyb. + screen)	43	36–52	53	45–61
Typing ($n = 78$)	47	37–56	57	45–68
Traditional office work ($n = 55$)	47	41–57	52	37–62

From the results it can be seen that the employees at workplaces with movable terminals prefer visual distances of 45 to 80 cm. Contrary to this, visual distances at workplaces with built-in terminals (conversational terminal B) are on average shorter by 15 to 20 cm. These comparisons give rise to the assumption that the employees make use of the mobility of the terminals to adjust individually preferred visual distances.

Body postures

Figure 4 shows the angles of various parts of the body measured during work. The results give rise to the following observations:

(1) Turning of the head (towards the documents) is more pronounced with

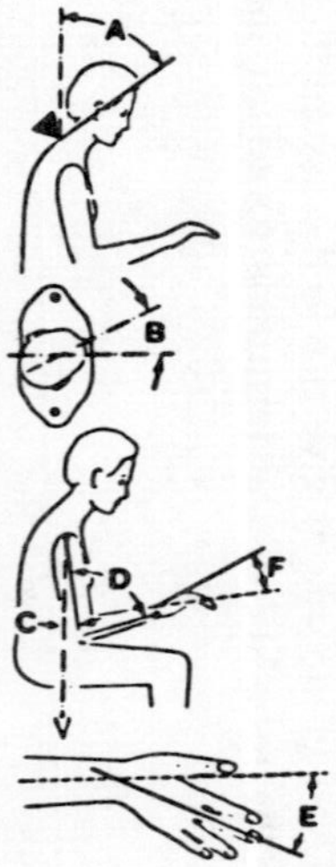

Figure 4. Assessed body angles of keyboard operators (x = mean values, s = standard deviation)

	Data-entry terminal (n = 53)		Conversational terminal A (n = 55)		Conversational terminal B (n = 54)		Typewriter (n = 78)	
	$\bar{x}$	s	$\bar{x}$	s	$\bar{x}$	s	$\bar{x}$	s
A = neck and head angle	59	8	61	11	55	9	58	7
B = lateral head rotation	8	5	12	11	11	9	19	11
C = abduction of upper arm right	22	11	32	13	18	7	16	7
D = elbow angle right	87	11	102	21	89	10	75	10
E = ulnar abduction left	9	11	13	10	20	11	21	10
right	13	7	12	9	14	10	17	9
F = extension of hand right	15	12	16	13	10	7	12	9

typing work than at terminal workplaces, where the documents can frequently be placed in front of the keyboard or behind it.

(2) Lateral lifting (abduction) of the upper arm is quite marked with conversational terminal A; here the employees can very comfortably rest the arms on the table. For the same reason this group shows the largest elbow angle.

(3) Lateral bending of the hand (ulnar abduction) shows a marked individual scattering: most are in the range of 0° to 40°. According to tendency, the values at workplaces with movable keyboards are somewhat lower than with built-in models and typewriters.

Factor analysis

The results mentioned so far show many connections between workplace dimensions and body postures. Many significant correlation coefficients confirm the relationships. This state of affairs led us to submit the data to a factor analysis. Three groups of factors appeared, which are of particular importance for our study.

As an example we have singled out here the results of the factor analysis of the workplaces with conversational terminal A (with movable keyboard) and represented them in Table 6 and Figure 5.

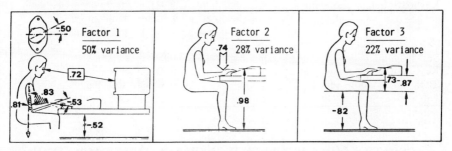

Figure 5. Factor analysis of body postures and workplace dimensions of conversational terminals (n = 55) with movable keyboard and screen. The figures show the factor loading

From this it appears that Factor 1 mainly contains characteristics of body posture; we are therefore calling Factor 1 'postural adaptation'. Factor 2 is principally limited to two characteristics referring to hand support, whereas Factor 3 comprises the characteristics of seat-level height. The results of this factor analysis can be interpreted as follows:

(1) The longer the visual distance, the more the upper arms are lifted and the more marked the stretching of the arms becomes, which also explains the decrease of head-turning and lateral bending of the hand.

(2) Smaller leg-room forces the employees to a slanting position of the thighs under the table, causing increased reclining of the trunk. This explains the increased visual distance with less leg-room, which again causes the other postures as described above.

(3) Factor 2 shows the relationship between keyboard height and frequency of resting hands and arms. The result is clear: the higher the keyboard, the more frequently the hands and arms are supported.

(4) Factor 3 trivially shows the correlations between seat-level heights and table heights above seat level.

Table 6. Factor analysis of the workplace data at conversational terminal A with movable keyboards and screens

Characteristics	F. loading	Factors
Visual distance	0.72	Factor 1
Upper arm abduction	0.81	————
Elbow angle	0.83	'adaptation of posture'
Ulnar abduction of hand	−0.53	50% variance
Knee room height	−0.52	
Turn of the head	−0.50	
Frequency of hand support	0.74	Factor 2
Keyboard height above floor	0.98	————
		'hand support'
		28% variance
Seat height	−0.82	Factor 3
Keyboard height above seat	0.73	————
Table height above seat	0.87	'seat height'
		22% variance

RELATIONSHIPS BETWEEN WORKPLACE AND COMPLAINTS

The analysis of the relationship between workplace dimensions and frequency of various complaints showed many significant correlations. In the following we would like to single out relationships which we consider relevant.

First, we examined the effects of table and keyboard heights, and came to a surprising conclusion: the lower the table and keyboard heights above the floor, the more frequently pains in shoulder, neck, and arms were indicated. To illustrate these findings we have chosen the example of the 'conversational terminal A group'; the results are shown in Figure 6.

In this connection, the evaluation of work height by the employees is of interest: about 80% of all employees at terminal workplaces judged their work heights as 'good'.

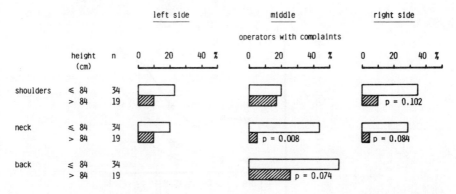

Figure 6. Incidence of daily and occasional pains with two different keyboard levels above floor at conversational terminal (A) (n = 53). One group is above and the other is below the median (84 cm)

The results concerning work heights were a great surprise to us, as they do not correspond to the actual recommendations of ergonomists. Observations at workplaces helped to clarify this state of affairs: at practically all workplaces the documents were placed on the table. No document-holders were used. That means the higher the table, the closer the documents are to the eyes. This condition led us to the following conclusions:

The higher the table height, the higher the documents, the better is the posture of head and trunk, and the fewer are the complaints.

Thus we cannot derive from our results a recommendation for high keyboard heights. We are rather of the opinion that the question of the best keyboard height above the floor should again be studied carefully with adequately positioned documents.

Another striking result of these analyses is the relationship between construction heights of keyboards and complaints: with data-entry terminals and conversational terminals A, keyboard heights above the table which are higher than the median values of 7–8 cm cause more pains in hands and arms. This result confirms the recommendation for low construction heights for keyboards.

RELATIONS BETWEEN BODY POSTURE AND PAINS

First, we looked into the question of the effect of the lateral bending of the hand (ulnar abduction) in operating keyboards. Medical findings show a marked dependence on the position of the hand. In Figure 7 the frequency of medical findings in the forearm muscles is set against the degree of ulnar abduction.

As Table 2 showed, clinical findings in the forearm muscles are not unusual. Figure 7 furthermore proves that the incidence of these clinical findings increases markedly when the ulnar abduction of the hands exceeds 20°.

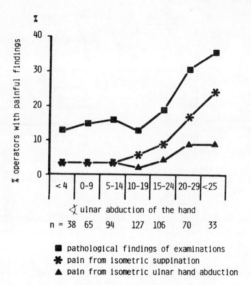

Figure 7. Percentage of medical findings in each group of ulnar abduction. The pathological findings of forearm examination include the pains caused by various hand movements (n = 271 all keyboard operators)

In Figure 8 the frequency of clinical findings of 'pains in the right forearm due to isometric contraction' is listed separately for an angle group of <20° and for another one of >20°.

The result is clear: this clinical finding is more frequent when ulnar abductions of more than 20° occur.

We also examined the question of possible relationships between resting hands and arms and complaints. We found many significant correlations between these characteristics in the data-entry terminals and conversational terminal A groups. We can state that, as a rule the 'hands and arms frequently supported' characteristic correlates with a lower incidence of pains in neck, shoulders, and arms.

Finally, we also studied the influence of the posture of the head on pains in the neck–shoulder area. The analysis, Figure 9 shows a certain relationship between the degree of head inclination or the degree of head-turning at the workplace and the frequency of clinical findings.

From this it follows that clinical findings in the neck–shoulder area tend to occur more frequently in all groups with increased head-inclination or turning of the head at the workplace, and the positioning of the documents significantly determine the posture of the head.

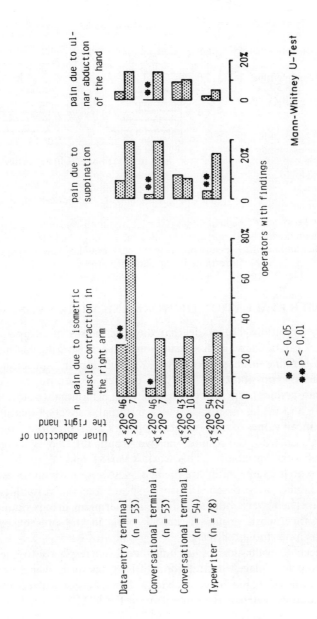

Figure 8. Incidence of medical findings in the right forearm of two groups with different degrees of ulnar abduction (p = significance of differences between the two angle groups)

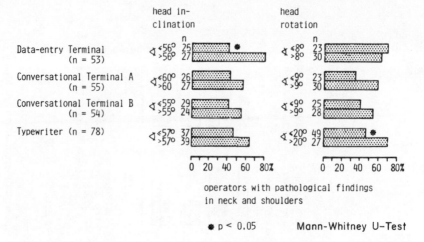

Figure 9. Incidence of medical findings in neck and shoulders in two groups with different angles of head posture The medical findings include limited head movability and painful pressure points in the neck muscles, as well as at tendon insertions in the shoulder area

SUBJECTIVE EVALUATION OF WORK AND WORK SATISFACTION

Occasionally objections are made that the true causes of complaints about terminal work can be found in dissatisfaction with new working conditions. This has led us to examine this hypothesis by means of a simplified questionnaire. In a bank we had the possibility of studying two groups doing the same work (payment transactions), but in one case without a terminal (traditional office work group), and in the other case with a terminal (conversational terminal B group). The mean values of the evaluation of work are represented in Figure 10.

As can be seen from the results, the two groups differ only as regards the evaluation of work variability. 'This work is varied' and 'With this work one always has to do the same thing' are the two questions on which the evaluations of work variability were based. Figure 10 shows that the variability of work clearly receives a lower evaluation from the terminal group than from the traditional office work group. Terminal work in the present example is considered as more monotonous.

This subjective evaluation of terminal work corresponds to the external observations of workplaces: in traditional office work the movements are walks to documents and office machines, exchange of information with colleagues, and telephone inquiries, whereas the same work at the VDT is limited to operating the terminal.

In order to obtain an assessment of work satisfaction the employees were also asked about their relationships with colleagues and superiors and the possibilities of making decisions and using their abilities. In replies to these questions there was no apprent difference between the two groups. The percentages of the 'satisfied' responses are in the range of 70% to 86%.

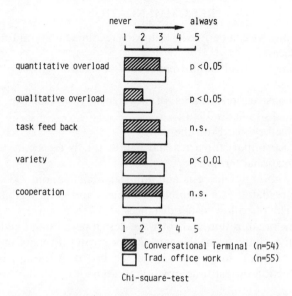

Figure 10. Mean values of perceived work situations in two groups with and without terminals but doing the same jobs (payment orders)

DISCUSSION AND CONCLUSIONS

Our field study shows that a number of employees at typewriter and terminal workplaces complain of pains in the neck–shoulder–arm area as well as in the hands. A comparison of employees carrying out payment transactions in banks shows that the use of terminals is associated with an increased incidence of complaints.

The observed complaints cannot only be considered as reversible fatigue symptoms; a thorough clinical examination has shown that about one third of the employees showed symptoms which are characteristic for pathological reactions after high muscle and tendon strain. In fact, many of the employees saw their doctor on account of such pains. There is thus an implied assumption of a causal relationship between constrained postures at terminals and other office machines and the impairments and symptoms described.

Measurements taken at the workplace show that unnatural body postures are frequent and are often caused by an unsuitable layout of the workplaces. Furthermore, extreme body postures are connected with increased complaints. The analysis shows direct connections between the dimensions and conditions of the workplaces, and body postures as well as complaints. These observations lead one to expect a decrease of complaints through correct design of workplaces.

RECOMMENDATIONS

From our studies we can derive the following recommendations for the design of VDT workplaces:

(1) Separate vertical adjustment for keyboard, display screen, and documents;
(2) Low construction heights of keyboards (for example, the German Standard stipulates 3 cm);
(3) Movable keyboards on the table;
(4) Space to support forearms and hands; this is especially important for conversational terminals;
(5) Seats with high backrests and adjustable inclination;
(6) Good readability of terminal characters and good legibility of source documents to allow visual distances of 40 to 80 cm;
(7) A work organization and software design permitting a reduction of the repetitive character of the work and a greater diversity of movements;
(8) Restriction of long periods of keyboard working time and the introduction of rest time in working hours.

REFERENCES

'Committee on Cervicobrachial Syndrome of JAIH' (1973). The report on the committee in 1972. *Jap. J. Ind. Health*, **15**, 304–11.
Hünting, W., Läubli, T., and Grandjean, E. (1980). 'Constrained postures of VDT-operators', in *Ergonomic Aspects of Visual Display Terminals*, ed. by E. Grandjean and E. Vigliani, Taylor & Francis, London, pp. 175–84.
Maeda, K. (1977). 'Occupational cervicobrachial disorders and its causative factors', *J. Human. Ergol.*, **6**, 193–202.
Maeda, K., Hünting, W., and Grandjean, E. (1980). 'Localized fatigue in accounting-machine operators', *J. of Occup. Med.* **22**, 810–16.
Martin, E., Ackermann, U., Udris, I., and Oegerli, K. (1980). *Monotonie in der Industrie*, Verlag Hans Huber, Berne/Stuttgart/Vienna.
Nakaseko, M. (1975). 'Thermographical study of skin temperature of fingers and hands on young female workers', Osaka City University, Japan.

Chapter 14

Occupational Stress Associated With Visual Display Unit Operation

COLIN MACKAY

Employment Medical Advisory Service, Health and Safety Executive, UK

and

TOM COX

Stress Research, Department of Psychology, University of Nottingham, UK

INTRODUCTION

Since the late 1960s an intense debate has developed around the whole question of industrial work and the demands associated with it. Fluctuating economic fortunes in Western industrialized nations, rapid organizational changes associated with the introduction of 'new technology' and the current and burgeoning concern with the social causes of illness have ensured that this debate has continued unabated.

It has become increasingly apparent that only a poor correlation exists between objective standards of living and measures of subjective satisfaction and happiness. This discrepancy has upset the traditional view that the quality of working life should continue to improve with increasing material well-being. Occupational stress has been advanced as a possible mediating factor in order to account for this discrepancy. An increasing number of the problems faced by the medical and social services appear to be caused, or exacerbated, by stress experienced at work, and it is increasingly seen as a threat not only to working life in particular, but also to the quality of life in general. Factors such as poor job performance, wasted leisure time, low job satisfaction, increased mortality and morbidity (perhaps resulting in premature retirement) and alcohol related problems have been shown to relate to the occurrence of occupational stress (Caplan *et al.*, 1975). These outcomes, it is suggested, reflect the cost of work: they balance out its benefits normally regarded as earnings, or personal and social development.

In parallel with general concern over psychosocial and health effects of industrial work has been a rapid trend towards the automation of work. The introduction of the new microelectronic technologies confer a number of advantages over existing systems, including greater control and precision resulting in improved quality and yield of products, reduced labour content leading to reduced costs and enhanced information flow resulting in greater organizational effectiveness. In the late 1950s and 1960 the over-riding, widely shared assessment of the social consequences of technological development equated the introduction of new technology systems with such organizational advantages, together with further benefits which may be conferred on the individual, related to his role in the work process. The advent of time-sharing facilities and the concept of man–computer interaction were consistent with these optimistic prognoses. Indeed, the term 'man–computer symbiosis' was coined to reflect the harmony between man and machine, the promise of enhanced information manipulation, an increase in decision-making capacity and a minimization in the routine aspects of existing work practices. The net result of these developments was to be an overall increase in the skill-level of users and operators of such systems.

With the development of computer technology and the growth of its application in a wide variety of industrial and commercial processes, there has been a corresponding growth in the use of visual display units (VDUs) and in business machines and equipment such as word processors in which they are incorporated. Following their introduction in the United Kingdom VDU users were almost exclusively computer personnel responsible for the design or operation of computer-based systems such as programmers and systems analysts. Over the last few years there has been a rapid proliferation of VDU-based systems in commerce and industry (Eason et al., 1974). At first their use was, on the whole, confined to specialized managerial functions, such as administration and accountancy or to research and development applications. Whilst these nonspecialist users faced some problems in adjusting to and coping with computer-based systems the advantages and benefits promised by new technology, in the main, have materialized. Since then, however, the overall direction and velocity with which VDU systems are being introduced has substantially altered. Thus one of the most rapidly growing forms of application is that in which the VDU is used to input data into the computer system: the so-called direct data entry task. Here, the VDU is used principally for the entry of information from source documents, which may be text, alphanumeric coded information from either printed or hand-written forms. This, and related types of task are usually of an intensive nature and are often carried out on a full-time basis. This has led to a new job grade known as 'VDU operator'.

Instead of the VDU being used as a tool to help the operator the person is really operating as a machine-minder. Perhaps the logical step to take here, since we seem to have reinvented the wheel, would be to get rid of VDUs and return to using pencil and paper again. However, opinion including that of the authors is in favour of their introduction but it should be ensured that they are introduced properly.

The widespread introduction of such systems, and the need for such jobs to be developed, has been associated with a number of problems, but the main concern has focussed on employment (Farrow, 1979) and possible effects on health (Mackay, 1980). Fundamental organizational change evokes feelings of apprehension at all levels from factory machinists to managers in offices. Uncertainty surrounding the need for one's own job in the new system or its possible restructuring with consequent changes in pay, promotion, and training etc. inevitably lead to anxiety. A number of health risks arising from the operation of VDUs have also been postulated. Much concern has been expressed over direct effects of VDU operation and microwave radiation induced capsular cataract and facial dermatitis were discussed in previous chapters. However, the current consensus view is that the majority of health related problems are a result of indirect factors. Typically, problems are manifested in reports of symptoms from operators related particularly to visual and postural complaints (Cakir *et al.*, 1980). Other aspects of the symptomatology may also be viewed as reflecting a general anxiety related to fatigue.

The majority of work directed towards the investigation of these complaints has been concerned with hardware and environmental factors, particularly the design of the VDU itself, the operators' workplace and visual surroundings. In comparison relatively little effort has been expended in investigating task-inherent demands of VDU operation and associated psychosocial factors in the operators' work environment.

The authors are particularly interested in discovering whether there are any short-term effects on the person's reduced well-being related to job satisfaction or, more seriously, whether long-term chronic health effects impair physical and mental well-being.

The remainder of this paper discusses briefly task-inherent demands in VDU operation together with relevant organizational factors as they relate to occupational stress. We will attempt to do this in two ways. First, by drawing upon existing information from the occupational stress literature which is relevant in examining the demands in VDU operation. Second, the chapter briefly reviews the existing studies of occupational stress in VDU operation. The final section of the chapter will suggest tentative solutions and highlight areas where further research effort is needed.

OCCUPATIONAL STRESS IN VDU OPERATION: ANALOGIES WITH REPETITIVE WORK

The authors have suggested that stress can be most adequately described as part of a complex and dynamic system of transactional processes between the individual and the environment (Cox and Mackay, 1979a; Mackay and Cox,

1978). Their model emphasizes that stress is a perceptual–cognitive phenomenon rooted in psychological processes. Essentially, the model states that stress will arise when the individual's needs and values (internal demands) are not balanced by environmental supplies from the work environment. Similarly, stress will arise when an individual's perception of his own abilities are not matched by the job demands being made upon him. It is important to note here that apart from the usual paradigm of mental stress (perceived demands exceeding perceived capability) the reverse situation of underload may also arise (perceived capability exceeding perceived demands), and it is this latter case which is of particular relevance for VDU operation. Underdemand is the principal characteristic of many repetitive work practices (Frankenhauser and Gardell, 1976), particularly machine-paced assembly lines (Cox and Mackay, 1979b).

Numerous studies, including those of the present authors (Cox and Mackay, 1981), have shown that workers engaged in repetitive tasks report the following:

(1) Underutilization of skills and abilities; the future relevance of these skills;
(2) Lack of autonomy and low levels of responsibility;
(3) Constraints, in particular machine-pacing and requirement for sustained attentional demand. (These second and third points combine to result in feelings of lack of control);
(4) Isolation/reduced social contact because of job constraints and environmental demands such as noise, or lack of proximity to others in the workplace.

Where simple clerical tasks have been computerized they often retain their monotonous and repetitive features characteristic of the above list. Moreover, in changing a job to suit the introduction of new technology an element of 'de-skilling' is often involved thus adding substantially to feelings of underutilization of pre-existing skills. In many individuals a considerable amount of time and effort will have been invested in learning and practising the skills required for the job prior to computerization.

The literature indicates that individuals engaged in repetitive work report feelings of boredom, dullness, and monotony, and lack of control over their work (Cox, 1980). Experimental and field studies of individuals engaged in repetitive work have indicated biochemical and physiological changes which may be early indications of the cost of such work and which may, if prolonged, lead to irreversible pathophysiological changes. However, there have been only a few studies carried out which have been able to demonstrate long-term health effects. Nevertheless, a cursory task-analysis of many existing VDU jobs indicates that they contain many of the undesirable features characteristic of short-cycled repetitive work, by failures to provide work which fit existing patterns of skill, and in so doing fail to meet the individual's need for challenging and interesting work.

OCCUPATIONAL STRESS IN VDU OPERATION: RECENT STUDIES

The handful of studies undertaken so far on occupational stress in VDU operators support the general points made earlier, which were drawn from the wider literature on the health effects of repetitive work practices. Thus in Gunnarsson and Östberg (1977) monotony experienced during VDU operation was clearly related to perceived feelings of lack of controllability and low levels of variety. Smith *et al.* (1980) report that rigid work procedures, high production standards and constant pressure for performance inherent in VDU operations were reflected in measures of self-reported stress and work demands which were substantially in excess of established norms (Caplan *et al.*, 1975). Their respondents also complained of negative effects on their emotional health as well as musculo-skeletal and visual problems. However, their data from other VDU sites suggest that perceived flexibility, autonomy and control over how work is to be carried out act as attenuating factors in the experience of stress. In these operators the greatest problems were those concerned with ambiguity over career development and future job activities.

One of the crucial variables in determining stress-related symptoms in VDU operators is linked with the perception of control by the system (or conversely, lack of control by the operator). Such control is evident in a number of ways. First, in some systems the processing power of the machine is exploited to such an extent that it is able to monitor the minute-to-minute keying performance of the operator. In many instances this information is used to determine levels of remuneration via piece-rate payment systems. Not unnaturally this level of control is often resented, is regarded with suspicion by many operators and, understandably, is associated with feelings of fatigue and stress. Second, very long response times from the computer, or those which are variable in length, create uncertainty and frustration. Third, technical disturbances and breakdowns, if they occur frequently, serve only to exacerbate these problems. All these factors substantially increase the mental load upon the operator and inevitably lead to fatigue. Thus Johannson (1979) and Johansson and Aronsson (1981) have shown that these aspects of machine control lead to marked psychoneuroendocrine mobilization in VDU operators as evidenced by increased urinary catecholamine levels. These effects do not go away when the person has stopped working but have a very insidious influence. They can affect one's ability to cope in situations where one has little control as well as leisure activities. There is a carry-over with these occupational demands.

The World Health Organization (1980), in their recent document 'Health Aspects of Well-being in Workplaces', makes a particular point of discussing the advent of the computerized system. They say that machine pacing, short-cycle work, and monotonous tasks, with little demand for the use of other than manual or muscular capacities, should be avoided. They suggest that we should be encouraging techniques for the development of human qualities and the use of decision-making capabilities and responsibility.

VDU task design must therefore seek to minimize repetitive elements in the operator's task by introducing variability in workload throughout the day, instead of long periods of concentrated work, whilst ensuring that the load is predicatable. This should be coupled with job-design features which allow the individual to have some discretion in how work is allocated over work periods and, by so doing introduce feelings of personal control and cater for individual differences in the need for brief pauses in work. Apart from these quantitative aspects of the VDU task, the qualitative features should also be examined. All too often data entry tasks require only the use of simple psychomotor skills, where only minimal exercise of intellectual abilities is possible. Thus work should be designed to be mentally challenging, but within the scope of individual operator abilities. Ideally, the shift should be towards the use of the VDU as a tool for carrying out a much larger enriched job, rather than regarding the VDU user as solely a machine operator.

When the operator's job consists of a variety of tasks, some of which may be related to VDU operation, minor faults in the ergonomics of the equipment may not be crucial. Conversely, when intense and continuous operation is required, the need for optimum workplace and screen characteristics becomes critical. However, solutions based entirely upon attention to ergonomic factors are not a panacea for low motivation and poor morale; work design and organizational factors are likely to be more important in determining the operator's overall acceptability of the computer. Thus in determining the acceptability of the system and minimizing possible indirect health effects we would link work design and organizational factors with primary prevention.

The switch from orthodox office handling routines to VDU-based systems provides an opportunity for job-enrichment and enlargement. Job-flexibility and some control over work-allocation should enable the promotion and utilization of individual skills. Whilst these are ideal outcomes stemming from the introduction of a computer system, they are only achieved by careful and thoughtful planning and implementation. The approach which appears to offer the most advantages and the one most likely to achieve these objectives is that in which designers and users of the system, particularly VDU operators themselves, are jointly involved throughout the various phases of design and implementation.

We are beginning to understand something of the psychological demands inherent in VDU operation. We are not yet at a stage where very detailed recommendations can be given in order to optimize VDU task design, partly because of incomplete knowledge of possible health effects, either in the short or long term, and partly because what existing knowledge is available has been slow to be implemented.

REFERENCES

Cakir, A., Hart, D.J., and Stewart, T.F.M. (1980). *Visual Display Terminals*, John Wiley, Chichester.

Caplan, R.D., Cobb, S., French, J.R.P. J., Van Harrison, R., and Pinneau, S.R. J. (1975). 'Job demands and worker health', *NIOSH Research Report, HEW Publication 75-160*, US Dept of Health Education and Welfare, Washington.

Cox, T. (1980). 'Repetitive work', in *Current Concerns in Occupational Stress*, ed. by C.L. Cooper and R. Payne, John Wiley, Chichester.

Cox, T., and Mackay, C.J. (1979a). 'The impact of repetitive work', in *Satisfaction in Work Design*, ed. by R. Sell and P. Shipley, Taylor & Francis, London.

Cox, T., and Mackay, C.J. (1979b). 'Introductory remarks: occupational stress and the quality of working life', in *Response to Stress: Occupational Aspects*, ed. by C.J. Mackay and T. Cox, IPC Science and Technology Press, Guildford.

Cox T., and Mackay, C.J. (1981). 'A transactional approach to occupational stress', in *Stress, Productivity and Work Design*, ed. by N.J. Corlett and J.E. Richardson, John Wiley, Chichester.

Eason, K.D., Damodaran, L., and Stewart, T.F.M. (1974). *MICA Survey*, Report of a survey of man–computer interactions in commercial applications, SSRC Project Report on Grant No. HR 1844/1.

Farrow, H.F. (1979). *Computerisation Guidelines: Guidelines for Managers, other Employees and Trade Unions Involved*, National Computing Centre Publications, Manchester.

Frankenhaeuser, M., and Gardell, B. (1976). 'Underload and overload in working life: outline of a multidisciplinary approach', *J. Human Stress*, **2**, 35–46.

Gunnarsson, E., and Östberg, O. (1977). 'The physical and psychological working environment in a terminal-based computer storage and retrieval system', *Report No 35*, National Board of Occupational Safety and Health, Stockholm.

Johansson, G. (1979). 'Psychoneuroendocrine reactions to mechanised and computerised work routines', in *Response to Stress: Occupational Aspects*, ed. by C.J. Mackay and T. Cox.

Johansson, G., and Aronsson, G. (1981). 'Stress reactions in computerised administrative work', *Supplement 50*, Reports from the Department of Psychology, The University of Stockholm.

Mackay, C.J. and Cox, T. (1978). 'Occupational Stress', in *Stress*, ed. by T. Cox, Macmillan, London.

Mackay, C.J. (1980). 'Human factors aspects of visual display unit operation', *Health and Safety Executive Research Paper No 10*. HMSO, London.

Smith, M.J., Stammerjohn, L.W., Cohen, G.F., and Lalich, N.R. (1980). 'Job stress in video display operations', in *Ergonomic Aspects of Visual Display Terminals*, ed. by E. Grandjean and E. Vigliari, Taylor & Francis, London.

World Health Organization (1980). 'Health aspects of well-being in workplaces', *EURO Reports and Studies 31*, World Health Organization, Copenhagen.

The second author gratefully acknowledges the support of the Medical Research Council and the US Army Research Institute (European Research Office). The opinions expressed are the views of the authors and do not necessarily reflect those of the Health and Safety Executive nor of either supporting body.

Chapter 15

General Discussion

Professor Shackel started the discussion by asking whether we should try to enrich boring but necessary jobs in the punch room and data room or spend more time concentrating on the electronic office of the future. A national official of a trade union replied that unless we reverse the whole process towards automation, work specialization and production lines we have to accept that there will be a concentration of routine, boring tasks such as VDU data input. This allows management to play with their design computers while others simply input the facts. We do not have to accept this an as inevitable process; we could offset a loss of productivity in a purely automated form against a gain in productivity when people are enjoying their work. This has been done successfully in Sweden, where a group, through a government agency, went into factories where workers were threatened with computerization. They analysed the work and, in three out of the four places, instead of workers remaining at their specialized, repetitive tasks, they shared out the work so that each had a greater variety. Perhaps each worker was marginally less efficient in doing a variety of tasks than he had been in doing one task. However, we know that it is difficult to maintain optimum efficiency throughout the day when doing a boring job. Should we go for maximum profits, introduce standardization, and use people as if they were robots? Is that fair to people? Perhaps profits and computerization should be used for introducing more interesting, more enjoyable, and therefore more efficient work methods.

Tom Stewart agreed that it was dangerous to assume that labour-intensive automated methods were more productive. If one is really concerned with productivity, other methods often lead to better quality work and the ability to get more out of staff resources. He had a client who, on two sites, had VDU-based punch rooms. Jobs there involved high-volume data entry, one-handed keying at 12 000 keystrokes an hour. In one location the operators were allowed to select their own jobs. They had the chance to walk about and chat between jobs; the trainees were on one rate, the average experienced workers on another, and the one or two experts had an additional rate. In the other location all operators were on piece-work. Every keystroke was counted and monitored. There were allowances for distractions and machine breakdowns, but everything was carefully calculated. The productivity for both locations, which used the same documents and did the same job, was identical. However, in the location with piece work they also had to have a staff of six people and a computer system, which ran for twenty-four hours every month, to analyse allowances, bonuses, corrections, etc. In economic terms the more human, flexible system was cheaper in obtaining the same productivity.

A group production consultant from a commercial company asked how monotonous tasks could be avoided, in practice, without using robotics. Colin Mackay said that it was particularly important in VDU operation not to have the operator working on it all day. We must try to dilute the repetitive aspects and give the operator something else to do, after they have had a block of direct data entry. Operators should also be given some control over the organization of repetitive and non-repetitive tasks. This would be a good start to reducing the problem.

Another questionner thought Mackay's presentation one-sided and suggested that VDUs have been, and are being, introduced into companies to enhance efficiency and therefore profitability. Consequently he thought it was totally unrealistic to make pious pleas for using the VDU in part of a larger job rather than for just data entry etc. Someone has to do these routine tasks.

Colin Mackay replied that if one continues to expect this type of work then operators are certainly going to experience the types of problems, that he had discussed. In reply to the suggestion that the dedicated VDU operator of today is only the punch operator of yesterday, Colin Mackay said that the same problems arose with those systems. He was frequently contacted by operators who suffer from problems arising from the repetitive nature of their work and lack of control. There were some sites where people were happy but there were others where the introduction of the system had followed this very rigid repetitive pattern creating these problems. In order to improve well-being and to minimize health problems the work must be structured so that a person is not doing the same task eight hours a day, five days a week.

A manager from a commercial organization was disturbed that the discussion was relatively narrow in terms of physical and mechanical solutions to something that goes much deeper. We should be exploring these avenues rather than merely providing varied work in order to reduce these stresses. Colin Mackay agreeed, but pointed out that the original idea of introducing new technology and VDUs into offices was that they should act as aids. However, we seemed to have gone off at a tangent and were merely replacing one kind of repetitive clerical work with another.

A safety officer from a computer supplier suggested that much of the monotonous work of several years ago had now been eradicated through the advent of computers. We should not get too involved with data-entry systems, as the work styles and duration of that particular type of occupation had also been shortened by the computer. Different input methods are obviously being developed all the time and he thought VDU entry, using keyboards, would not continue for the next twenty years. His company and most other suppliers offer to advise users on how they should set up their equipment, but they were very reluctant to force such information on users, as many prefer to make their own decisions; indeed, many do not want to be told what to do and where to put equipment as very often they are very restricted in their choice.

A factory medical officer from a commercial company asked Hünting if his findings depended upon the length of time that an operator had worked with a VDU. If someone had worked at a VDU for several months their muscles would become accustomed to the movements so that the pain would disappear. Pain is always a very difficult symptom to measure, and what is painful to one person may not be painful to another. Hünting replied that he only took subjects who had been working with VDUs for over six months. Professor Wagenhäuser (Rheumaklinik of the University Hospital Zürich) helped with the investigation and all the findings were made by a medical doctor. Of course, there are significant individual differences in tolerance to pain but at the data entry terminals the work load and the incidence of impairment were both very high.

In reply to a consultant from a computer supplier Hünting confirmed that he had included the height of the screen in relation to the operator's line of sight in his study and a relationship was found. The more forward the functional angle of the neck and head the higher was the incidence of impairment. He did not conclude with a recommended optimum visual angle as this depended upon the individual's posture. The more one has to look down on the display the worse the neck pains are going to be. Hünting added that he found no adaptable document stands in his field studies. These studies were carried out in the payment departments of banks and documents were always placed on the tables.

Tom Stewart fully supported the view that a low screen leads to neck and shoulder problems but warned that one must not imply from that that 'the higher, the better', because, equally if the screen is too high, holding one's head up is very tiring. We are not accustomed to sitting upright with our eyes focused horizontally; we all naturally slump forward slightly and find it more comfortable to look slightly down. The optimum position is not the higher, the better (or the lower, the better) but should be in a middle range which one has to define.

R. A. Weale of Reuters and Professor of Visual Sciences at the Institute of Ophthalmology said that he and his colleagues provide the data that are being grossly misused. He wondered whether the attempted quantifications of what regrettably appeared to be a pseudo-science could be justified. In *Ergonomic Aspects of Visual Display Terminals*, edited by Grandjean and Vigliani, there is a chapter which complains of the pressure that scientists are under to produce data to be used by ergonomists, particularly those concerned with VDUs. This had been apparent in the discussion and the earlier example was by no means isolated. In many cases when data are presented, no controls or only inadequate controls are quoted or obtained; in other instances, for example, radiation, controls have been obtained but are ignored. In one trade union people over forty-five are virtually prohibited from using VDUs. One can, in fact, advance sound visual reasons why people of this age will have fewer problems with VDUs than younger people. There is a chaos of data which everyone is trying to come to terms with. It would be simpler to call it a day, and say that we are going to be humane and inefficient or humane and efficient, and ignore the data.

Professor Shackel said that there were many statements in the previous general discussion pointing out that control data cast great doubts upon the thesis on cataracts (Chapter 9), and that it is the concern of ergonomists and other human scientists to have valid data from controlled experiments. Mr Hünting's examples were of valid observations. He compared VDU operators with typists and office workers, who acted as controls. This method produced differential results, not just a statement that VDU operators are at a severe disadvantage. He hoped that there would be many more such studies.

A data processing manager from a commercial company asked Peter Stone 1) if the eye tests that he described could be carried out by any standard optician or whether special techniques or equipment are required and 2) how any natural light in offices could be dealt with. Stone thought that the majority of the tests, especially the screening tests, could be carried out by most opticians. Unfortunately some of the complexities of binocular co-ordination are not understood by everyone in the ophthalmic profession, therefore in some circumstances it may be necessary to seek the advice of a specialist. High Street opticians may be interested in industrial problems but he stressed that the tests should be done in conjunction with the workplace. The Association of Optical Practioners are interested in industrial welfare and many of their members go into industry and study the person at the workplace. A simple spectacle prescription from an optician may not be the answer to the problem of eye strain at work.

The question about natural lighting was an issue not only applicable to VDU work stations but to many others. An obvious answer was to provide blinds but the window has functions other than just admitting light. It also provides a view, which is extremely important. Some sort of screening is therefore necessary which will not totally obliterate the light. One could use horizontal screens which shut off a substantial amount of light but do not entirely shut off the view, or, one of his favourite methods, net curtains. Net curtains sound Victorian but have great advantages in diffusing light efficiently and reducing glare. They will not, however, reduce specular reflections. If the person has to sit beside a window or there are specular reflections in the screen, blinds or matt non-gloss curtains will solve the problem. To eliminate specular reflections by avoiding the 'offending zone' the screen can be turned to a different aspect, although the operator should not be facing the window as glare problems will then occur.

Tom Stewart said that there were various solar-control films which work quite well in some cases. There were many solutions to this problem but one needed to find the one that suits the particular situation. The net curtain has, in fact, been upgraded. There is a German roller-blind of a net material which could be bought in different densities with different apertures depending upon the amount of light which needs to be shut out. He emphasized Peter Stone's point that the view from a window serves a very strong psychological purpose and also enables one to relax the eyes. However, he would say to architects that we do not need windows that are the size of the entire wall to get a nice view. You can get a remarkably good view from a small window with none of the enormous heating and lighting problems encountered with picture windows.

A training and safety manager said that his company was currently in the middle of its draft policy on VDU operation and he was rather concerned that a major insurance company and ASTMS had agreed to a policy, which included a free eye test, and were also prepared to purchase any spectacles deemed necessary as a result. His company had consulted three members of the Association of Optical Practitioners who had all said that there was no reason for prescribing spectacles for VDU use if due attention was paid to the ergonomics of the work situation. Peter Stone said that it is well known that some people need different optical prescriptions according to the task; one for reading and another for driving, for example. Consequently, when looking at the viewing distance for a VDU there may be a few cases where a prescription is necessary. Sometimes a special investigation for a particular person at that particular distance is necessary. However, Stone did not think that the provider of the spectacles should necessarily be the employer, as the general provision of spectacles seems to be fully covered by the National Health Service. The core of the problem may lay with the opticians, who perhaps do not really appreciate the details of a specific working situation.

A director of a company supplying vision screening equipment said that Colin Mackay stated at the end of his chapter that, because incomplete knowledge of possible health effects still exist, he was not able to make any very detailed recommendations. This does not conform with the suggested guidelines and recommendations laid down in Colin Mackay's *Research Paper No. 10*. The director was a member of the Association of Optical Practitioners and his association did not at this stage agree with all those recommendations. It was not logical to say there is no requirement for eye examination unless, after some months, people still have headaches. Peter Stone and others referred to these task problems, and Wilhelm Hünting mentioned impairment of visual performance. Although he might be accused of having a vested interest, the director said that his concern was quite genuine in this matter. If one expects people to use their eyes then one ought to ensure that those eyes are usable. Many years ago he conducted optical tests within industry and found that some 25% of the working population had visual deficiencies of which they were were unaware. It would not be a bad idea to ensure that if you had to use your eyes they were functioning properly.

A data capture manager from a commercial company asked what had happened to the responsibility of the individual? Why was it the responsibility of a company to ensure someone's eyes are all right? He suggested that it was not the responsibility of the major companies in the UK to provide these services.

Colin Mackay said that in his research paper he was trying to stress the importance of ergonomics. If the design is correct then any residual problems with an individual's eyesight can then be corrected by specialist advice.

Peter Stone thought that there was an important percentage of people who were either under-prescribed or have not had a prescription for spectacles. Where visual work is concerned it is clearly important to remedy that. In the present climate, with so many complaints about eye strain and VDUs, it would seem that pre-employment screening checks are advisable for the next few years. These need not include involved optical examinations. Pre-employment visual screening should be carried out in much the same way that one would conduct pre-employment audiometry before admitting a person to a noisy industry.

S. McKechnie, of ASTMS, said that during the course of discussions that the union had had with the Association of Optical Practitioners it was clear that the Association had given much thought on how the screening systems should work. They had devised a form for the employer to fill in when referring someone to an optician and another for the optician to return to the employer. Under the National Health Service everyone has the right to a full eye examination every two years, and the Association's view seems to be that they do not want that right to be undermined. Under the proposal to have mass testing in companies, with a charge being made for a National Health examination, the worker's right to visit an optician of their choice is undermined. The Association was about to bring out their own document on this subject. In view of the discussions and controversy ASTMS was currently issuing a very detailed statement on their proposals for eye tests. She agreed with Colin Mackay that the visual standard required for VDU operation need not be higher than for other types of clerical work. However, she thought Colin Mackay's subsequent arguments that eye tests are not needed were wrong. She also thought that the union with the policy concerning the exclusion of forty-five-year-olds had been so condemned by other unions that she was sure that they would soon withdraw their statement. It is a very wrong approach to choose a safe worker rather than a safe place to work.

A medical officer from a commercial company said that one important factor had been omitted from the discussion on eye-screening. Most large employers have occupational health staff and, as part of the routine pre-employment, medical examinations visual screening is carried out. This may not be to the extent that one would desire for specialist operations but routine screening would detect any deficiency in visual acuity which was likely to affect an operator on his first appearance in front of a screen. If one accepts the premise that VDU operation is likely to cause eye strain and that stress symptoms may also arise then there was a case for some form of screening of a person's mental health, mental stability, and reactions to stress before they were subjected to the operation of a VDU.

Tom Stewart thought that someone with defective vision would normally expect to obtain prescribed spectacles from the National Health Service, who would presumably pay for them. However, if an employer creates or changes a job so that it puts new and different demands upon the individual then the employer should be responsible for the change by paying for new spectacles if they are required. Preferably the job should be designed so that it does not make unacceptable demands of most people. Most VDU jobs can be designed so that people with relatively normal vision can use them. This seems a much better approach than selecting people who have the requirements that fit the equipment. We should design equipment so that normal people can use it in a normal way. We should try to adapt computers to people, rather than select people who must work on computers the way we currently design them.

A national official of a trade union said that many people think that offices should not be humidified because of the danger of 'Legionnaires' disease', but many libraries and art galleries do use water-washed air conditioning to prevent desiccation of books, etc. Presumably these places would not have the kind of skin problems, conjunctivitis, and drying of the cornea that may cause eye strain in overheated, overdried offices, and she wondered if these water-maintained viruses were really a problem.

S. McKechnie of ASTMS said that most of the papers on this air-conditioning problem are American. The illness is referred to as 'humidifier fever' and is rather like 'flu. There are not many water-purified systems in the UK but some air-conditioning firms are very concerned about the problem. The DHSS has issued a circular after an outbreak of Legionnaires' disease in a hospital. It is essential with any water-purifying system already installed that there are regular changes of the water, and that the temperature and chlorination are sufficient to ensure that these bacteria cannot survive. This did not seem to be an insoluble or particularly difficult problem.

However, another point concerned her women members, many of whom strongly believe that the level of miscarriage among VDU operators is higher than in other jobs. Initially it was thought that radiation could be the cause but there was no good evidence to support that at present. Another theory suggests that it may be linked with the growing levels of stress, particularly where the VDU is linked to a mainframe. Another suggestion was that the fixed posture of VDU work puts stresses on a woman's body that are very inimical to a succesful pregnancy, particularly in the early stages. The other explanation put forward was that increasing discussion of the problems of pregnancy makes it seem as if there are more problems. Changing social values and attitudes have resulted in more open discussions on such matters and have given the appearance of new problems.

Colin Mackay thought that the Toronto Star studies did suggest that radiation was not the cause. As with Zaret's work, we do not know the baseline levels of miscarriage. Although it would be a great leap in the dark, one could in theory suggest a link between very high stress levels, neuroendocrine changes, and detrimental effects upon the foetus.

A safety officer from an insurance company wondered whether the Health and Safety Executive could not do more to compel manufacturers into giving information on VDU equipment. Section 6 of the Health and Safety at Work Act does place a duty on manufacturers, designers, and importers not only to give such information but also to carry out research, tests, and examinations so that they know what is involved in their equipment. Brian Pearce had stated that some manufacturers did not know the luminance of their screens: this was illegal.

An occupational hygienist from a commercial company suggested, first, that the enforcing authority for areas where most VDUs are used will, in fact, be the local authority and not the Health and Safety Executive, and second, that the Health and Safety at Work Act does concern itself with VDUs, as its aim is to ensure health, safety and welfare.

Brian Pearce said that while speaking elsewhere he had showed a slide as an example of the 'New Technology Sweatshops'. This was an American photograph of rows of VDU operators at work. In the audience was a member of the Health and Safety Executive who said that in the UK such a situation would have some form of closure or improvement order placed upon it. When asked, he could not, however, tell of any VDU installation where the Executive had actually taken such action. The Health and Safety at Work Act did not have any effect in relation to the common problems under discussion. It may be applicable to VDU installations, but it had never been used in that way.

Another delegate said that he worked for a company which was doing everything possible to improve the environmental aspects of VDU operation. However, it was limited because neither the union nor the company were getting the required information from the manufacturers, creating problems for both sides, and other companies must be in exactly the same position.

S. McKechnie of ASTMS accepted the criticism that if one aims for specific recommendations, technology is likely to overtake them within a year, but when most of the unions were drafting their guidelines they were dealing with very poor images on VDU screens. How then, when sitting at the negotiating table, could both sides agree on what actually is a 'clear stable image'? Unless recommendations are made more specific, one only transfers the argument to another area. It is not sufficient to write into agreements that there will be a clear, stable image if there is no mechanism to decide what that is.

Brian Pearce agreed that one would increase the possibility for further discussion and dispute if there were no specific recommendations, as each person would have their own definition of a clear, stable image. Therefore the solution is not to get more technical with more numeric quantities but to have a human performance-based specification. With a standard for human performance-based user legibility and readability of VDUs there would be a point on which management and unions agree. Manufacturers would submit their VDUs to an independent test to determine the clarity and stability of their displays.

Professor Shackel wondered if there was a need for a VDU evaluation service which would not have the force of law: a service to be performed by an independent, neutral body on behalf of users and suppliers.

He appreciated the problems involved in this, which was why he was somewhat hesitant. However, he continually received messages that such a service would be useful.

Section 3

Chapter 16

Optimal Presentation Mode and Colours of Symbols on VDUs

GERALD W. RADL

Overath-Marialinden, Federal Republic of Germany

PRESENT STATUS IN APPLICATION OF VDUs IN THE FRG

In 1981 approximately 300 000 VDUs were in use in the FRG. The category of VDU means CRT-displays for presentation of numeric data and textual information. Approximtely 80 000 of these VDUs are used, mainly continuously during the daily working shift, for data input into electronic data-processing systems. Approximately 150 000 are employed with computer systems and are in use either continuously or intermittently for dialogue with technical information-processing systems. Approximately 30 000 units are installed in offices for word-processing, and a small number of them are in use for text communication. Approximately 40 000 VDUs are used in industry and in research and development areas. If the analogue television monitor is also included within the definition of a VDU, 300 000 or more television screens must be added, which are installed in industrial television systems.

PROBLEMS AT WORKPLACES WITH VDUs

The image of work with VDUs is generally bad in the FRG. The causes are complex, but it is possible to point out the following main factors, which are caused as much by hardware ergonomics and software design as by the change of the work content and organizational conditions for many employees.

To date, some of the screens and keyboards are badly designed. The most unsatisfactory points are low luminescence level on the display, too large contrast between the screen area and the other areas within the visual field at the workplace, too low contrast between characters and background, flicker of the display, reflections on the screen, and the design of the whole unit such that it is often impossible to use ergonomically. Relatively poor workplace design and bad positioning of the screen, the keyboard and other work aids are causing unsuitable working positions and muscular tension during work.

The illumination conditions at many VDU workplaces are unsatisfactory. The German standard prescribes 500 lux for office workplaces in conventional rooms and 700 lux in office landscapes (DIN 5035). But these are only general recommendations to avoid glare and reflection. The existing illumination problems are caused by the relationship between relatively low light densities and the daylight or the artificial lighting conditions at the workplaces.

Eye defects are often the reason for an increase in the workload of many operators working with VDUs. However, these defects are not caused by VDU use. Field studies have shown that more than 50% of all West Germans have non-corrected eye defects (Hartmann, 1977).

In some cases the use of VDUs has induced an increase in information transmission rates between man and the technical information-processing systems (Cakir et al., 1977, 1980), thus increasing the mental load.

VDUs have also become symbols of anxiety for employees about technical and organizational changes in white-collar jobs, mass unemployment by increased use of technology, de-skilling and increased control by the computer (Radl et al., 1980). I think that we will solve all the technical problems of adapting VDUs to man within the next few years, but the psychological and social questions in this field will become increasingly important.

ERGONOMIC REQUIREMENTS

The role of ergonomics

Ergonomics cannot solve existing problems in working life with a few ingenious measures. All our experience has shown that questions of work physiology as well as the psychological problems related to the use of new information processing technologies are diverse and the solutions complex. Ergonomics seems to be a puzzle. Many small steps have to be taken before a working system is fully adapted to man. Our experimental results and practical experiences with the 'positive presentation mode' and coloured symbols on the screen may be such small steps.

Ergonomic requirements for the VDU

According to the present status of work physiology and ergonomics it is possible to design equipment and workplaces for information processing by using VDUs so that the stresses arising through posture and the optical information input system are no greater than those of well-designed conventional office workplaces without VDUs. Through the conscious application of ergonomic knowledge it is possible to reduce the work demands on the employee carrying out information-processing with VDUs to below the level of strain at similar but non-computerized work-stations without screens. But what aspects are of special importance in the ergonomic design of work-stations with VDUs? What requirements does ergonomic research and practical experience place on these work-stations? Listed below are the specifications with ergonomic relevance for which the manufacturer of the technical equipment is responsible:

(1) A high level of screen luminance.
(2) An optimal level of contrast between characters and background on the screen under all illumination conditions where the work-station is in use.
(3) In cases where colour is used on the screen, the selection of colours which do not reduce the level of perceived luminance and which do not cause colour-misperceptions.
(4) A minimum of flicker on the screen.
(5) A minimum of direct light reflectance at the surface of the screen.
(6) Easily readable characters of optimal size, especially a high dot-matrix, and significant forms.
(7) User-oriented design of the program routines and of the whole software interface, which is relevant to the operator.
(8) Avoiding X-ray emissions from the tube.
(9) A keyboard which is separated from the screen and which can be easily adapted for use by people with different body dimensions in physiologically favourable sitting and writing positions.
(10) An optimal visual design of the keys, improving the legibility of the lettering and avoiding light reflections.
(11) A design and positioning of keys which are adapted to the characteristics of man's biomechanic, motor, and tactile performances.
(12) An understandable manual, including not only all information necessary to solve problems under normal conditions but also the basic ergonomic facts of using equipment with VDUs.

EXPERIMENTAL INVESTIGATIONS OF THE 'PRESENTATION MODE'

The problem

We have registered 8000 to 25 000 eye movements between paper manuscripts and the screen at word-processing stations during an eight-hour working period (Radl, 1980). Different luminance levels on the screen and on the manuscript cause avoidable visual load, because the adaptation mechanism does not work within the short time of one eye movement cycle. Cakir et al. have found up to 35 000 eye movements between the screen and the paper manuscript under specific work conditions (Cakir *et al.*, 1977). The very important requirement of a high luminance presentation on the screen and equal luminance levels on the screen and in the other parts of the visual field of the workplace (i.e. on the paper manuscripts) seems to be more easily fulfilled when using the so called 'positive presentation', i.e. negative contrast (Bjørset and Brekke, 1980) or dark symbols on a light background. In addition, positive presentation seems to be an effective and ergonomically ideal method of avoiding direct light reflections on the screen. It is well known that reflections on the screen from lamps, windows, or light wall and ceiling surfaces quickly lead to disturbances of vision and also to asthenopic complaints (Cakir *et al.*, 1977; Östberg, 1977; Radl, 1981).

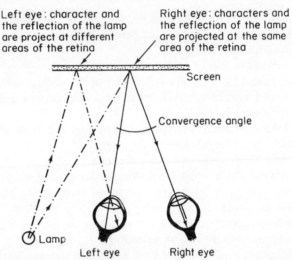

Figure 1. Disturbance of the convergence control systems
by reflections on the screen

Light reflections on the screen will, at a distance between 50 and 70 cm, be seen on different parts of the screen by both eyes or only be visible in one eye. In such a case one refers to a convergence disturbance, since it is not possible for the visual mechanism to emit the pictures received from both eyes.

In order to avoid, or reduce as far as possible, light reflections on the screen the following precautions should be taken:

(1) Positive presentation (the best method).

(2) Micro-mesh filter (the effect is good, but it decreases the luminance).

(3) Matting of the screen surface (the effect is very good, but the sharpness of the characters may decrease).

(4) Last but not least, consideration for the positioning of the VDU and a high-quality interior lighting system (e.g. mirror raster lamps with 'darklight' effect) can also greatly help towards avoiding reflections on the screen.

The experimental question

Can differences be found between positive and negative presentations even during a short working period with the VDU?

Method

The performances, error rates, and ratings of visual comfort were compared when individuals had to perform standard tasks which simulated the visual performance of data and text input to a VDU from a paper sheet. The same VDU was used for both presentation modes with a refresh rate of 66 Hz to avoid flicker when using positive presentation. The tasks were carried out under good illumination conditions (500 lux) at an ergonomically well designed workplace.

Subjects

The subjects were twenty-four males and females, all participants of an introductory course in text communication.

Results

The results are shown in Figure 2.

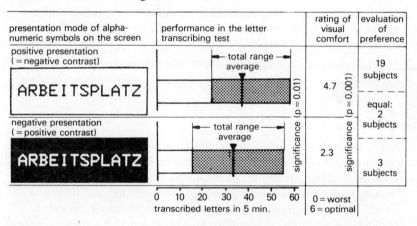

Figure 2. Performance in a letter-transcribing test and ratings of visual comfort and of preference under positive and negative presentation modes of the alphanumeric symbols on a VDU screen

The differences are significant, and can be interpreted on the basis of physiological optics. Positive presentations cause a higher average luminescence level on the screen, visual acuity and the depth of focus of the eye increase

(Hartmann 1977; Häusing, 1976), and too large brightness differences between the screen and the paper sheet are avoided.

Conclusions

Positive presentation (dark symbols on a light background) was found to be significantly better than negative presentation. Positive presentation can be recommended not only on the basis of the results of this experiment but also for the following reasons:

(1) The adaptation conditions for the eye are better, especially when eye movements between the VDU screen and a paper sheet occur often.

(2) In cases where it is impractical to avoid a bright optical environment the probability of discomfort glare is reduced.

(3) Positive presentation is also an effective and ergonomically ideal method for avoiding reflection on the screen. This can be demonstrated in many practical cases.

(4) Today VDUs with positive presentation are technically possible and can be produced at an acceptable cost.

Indeed, the German safety regulations for VDU workplaces in offices today are recommending a positive presentation mode on the screen (Verwaltungs-Berufsgenossenschaft, 1979).

EXPERIMENTAL INVESTIGATIONS OF COLOURS ON THE SCREEN

The problem

Colour is an important factor in our visual world and can be an aid for more effective use of man's visual perception system. It is important to differentiate between three various functions of colour on the VDU, namely:

(1) The use of one colour to adapt the physical characteristics of the light emitted by the tube to the human eye (mono-colour displays);

(2) The use of colours as a formatting aid on the screen (multi-colour displays); and

(3) The use of colours as a visual code (also on multi-colour displays).

We have investigated mono-colour displays and the performance of man by using colour as a visual code on a display. Krueger (1981) and Robertson (1980) have published the results of their investigations about the use of colour as a formatting aid.

Effects of coloured symbols on a VDU screen

The question

What symbol colour is best, when light symbols are presented on the dark background of the VDU screen?

Psycho-physical factors

The effect of the spectral sensitivity of the eye is well known (Hartmann, 1977). As Figure 3 shows, the brightness sensitivity of the human eye is greatest at a wavelength of 555 nm. Also known is the effect of the chromatic aberration of the optical system which can make the eye short-sighted for blue-violet up to 1.5 dioptres (Le Grand, 1957).

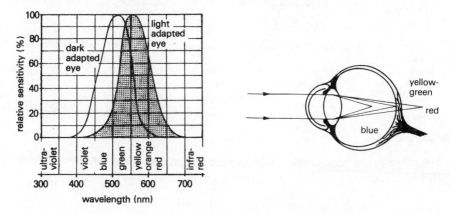

Figure 3. Spectral sensitivity and chromatic aberration

Experimental method

White and six different symbol colours were produced using CRTs with different phosphors and coloured Plexiglass filters. Figure 4 shows the colours used within the German Standard colour diagram (DIN Farbenkarte which is similar to the Munsell diagram). The different luminescences of the coloured signs were also measured. Under laboratory conditions and at a well designed workplace, illuminated with 500 lux without glare and without reflections on the screen, individuals had to carry out a task similar to those described in the first experiment. They were forced by the letter transcribing test to make eye movements between the screen and the well-illuminated paper sheet. The duration of the test run under each condition was 10 minutes. After the tests with all seven symbol colours the individuals had to compare each colour condition of the symbols with every other (twenty-one comparisons) by considering which they would prefer on the VDU at their workplace.

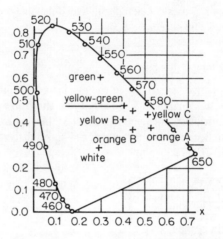

Figure 4. The symbol colours used within
the standard colour diagram

Subjects

The subjects were thirty males and females between nineteen and forty-two years, all with experience in VDU work and all without eye and colour defects.

Results

Figures 5 and 6 demonstrate some of the results.

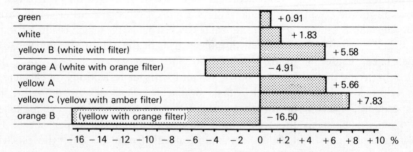

Figure 5. Relative performance in the letter-transcribing test related to the
individual's average performance under all colours

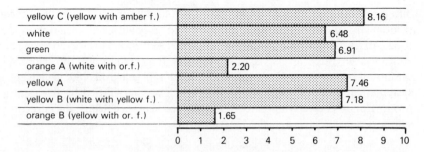

Figure 6. Preference scores of the different symbol colours. The scale ranged from 0 (no preference in all comparisons) to 10 (preferences in all comparisons of this colour with every other)

The investigated symbol colours produced only relatively small differences with respect to the readability, and not all differences are significant; but high and significant differences are found with respect to the ratings in the paired comparison test. The brightness and the contrast of the symbols on the VDU screen seem to be more important than the symbol colours when they are within the recommended area of the spectrum.

Conclusions

Coloured light-emitting phosphors to reduce the bandwidth of the spectrum (and also the luminescence) normally have no advantage. Only tubes with phosphors which transform the full energy of the electronic beam into a reduced band within the yellow-green region can be recommended. When such coloured symbols are preferred it seems to be caused mainly by psychological factors rather than by physiological mechanisms. This finding could also be verified by practical experiences since the investigation.

Discrimination of coloured symbols on different coloured backgrounds on a television screen

The question

What effects do different combinations of symbols and background colours have on the detection rates of small symbols, on the rate of correct colour discrimination, and on the reaction time?

Method

In this experimental study subjects were required to detect as quickly as possible and to name the colour of small moving squares on a television monitor with different coloured backgrounds. Five symbol colours and seven background colours (six colours as shown in Figure 6 and 'grey with 40% visual noise') were used. All trials involving a single background colour were presented in one block. Symbol size (three levels), symbol velocity (fast and slow), directions of the symbol movements and illumination level (10 lux and 480 lux) were also systematically varied. Observer performance was quantified by registering the proportion of correctly and incorrectly named symbol colours, the percentage of symbols not detected and the reaction time.

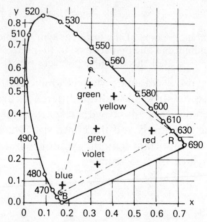

Figure 7. The colours used and the limitations of the colour TV screen within the standard colour diagram

Subjects

The subjects were thirty well-sighted young men without colour defects.

Results

Figure 8 shows that the various combinations of symbol/background colour produced error rates between 4% and 95%. The accuracy and speed of perception were also influenced by the symbol size and to a lesser extent by the symbol velocity and the direction of motion. The level of illumination had a comparatively small influence.

Conclusions

In view of the high error rates observed in this study, careful preliminary considerations are required for the colour-coding of symbols of television displays. The investigations by Krueger and by Robertson brought 'better' results for multi-colour displays: they used special tubes with a much higher dot-matrix than the colour television tube, which we have used in our experiment. They also used blue and red with a high white content.

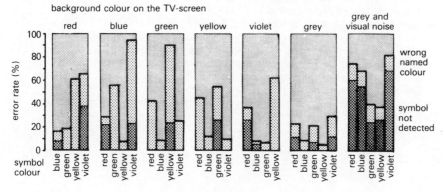

Figure 8. Error rate for the different symbol–background colour combinations

REFERENCES

Bjørset, H.H., and Brekke, B. (1980). 'The concept of contrast: a short note and a proposal', in *Ergonomic Aspects of Visual Display Terminals*, ed. by E. Grandjean and E. Vigliani, Taylor & Francis, London, pp. 13–24.

Cakir, A., Reuter, H.J., Schmude, L.V., and Armbruster, A. (1977). 'Untersuchungen zur Anpassung von Bildschirmarbeitsplätzen an die physische und psychische Funktionsweise des Menschen' (Investigations into the adaptation of VDU workstations to the physiological and psychological functions of man). Forschungsbericht reihe *Humanisierung des Arbeitslebens*, Band 1, Der Bundesminister für Arbeits und Sozialordnung, Referat für Press und Öffentlichkeitsarbeit, Bonn.

Cakir, A., Hart, D.J., and Stewart, T.F.M. (1980). *Visual Display Terminals*, John Wiley, Chichester.

DIN 5035: *Innenraumbeleuchtung mit künstlichem Licht* (German Standard: Indoor illumination with artificial light).

Hartmann, E. (1977). *Optimale Beleuchtung am Arbeitsplatz* (Optimal illumination of the workplace), Kiehl Verlag, Ludwigshafen.

Häusing, M. (1976). 'Colour coding of information on electronic displays', in *Proceedings of the 6th Congress of the International Ergonomic Association*, The Human Factors Society, Santa Monica, California, pp. 210–17.

Krueger, H. (1981). 'Farbige Darstellung von visueller Information' (Coloured presentation of visual information), Report, Institut für Arbeitsphysiologie der Technischen Universität, Munich.

Le Grand, Y. (1957). *Light, Colour and Vision*, John Wiley, New York.

Östberg, O. (1977). 'The physiology, psychology and measurement of glare', *Technical Report 24 T*. Department of Work Science, University of Luleå, Sweden.

Radl, G.W. (1980). 'Ergonomie, Arbeitsplätze mit Bildschirmgeräten' (Ergonomics, Workplaces with VDUs), *Report No. 128011*, Nixdorf Computer AG, Paderborn.

Radl, G.W. (1981). 'Ergonomische und arbeitspsychologische Empfehlungen für die Textverarbeitung mit Bildschirmterminals' (Ergonomic and psychological recommendations for word processing with VDUs). Akzente, Studiengemeinschaft *Akzeptanz neuer Bürotechnologien*, Seeshaupt.

Radl, G.W., Reichwald, R., Scheloske, G., Schreiber, R., and Weltz, F. (1980). Akzeptanz neuer Bürotechnologien, Bedingungen für eine sinnvolle Gestaltung von Arbeitsplatz, Organisationsstruktur und Mitarbeiterbeteiligung (Acceptance of new office technologies. Conditions for intelligent design of workplace, organizational structure and participation of the employees), Akzente, Studiengemeinschaft *Akzeptanz neuer Bürotechnologien*, Dusseldorf.

Robertson, P.J. (1980). 'A guide to using colour on alphanumeric displays', *IBM Technical report*, White Plains NY.

Chapter 17

The Possible Benefit of Negative-Ion Generators

L. H. HAWKINS

Department of Human Biology and Health,
University of Surrey, UK

INTRODUCTION

I would like to deal with the question of ionization and its possible influence on the way people work, behave, and feel in a working situation. It is a subject not without its critics and sceptics. Nevertheless, levels of ionization in the modern working environment are markedly different from outdoor values and it is very important that we should at least understand whether these changes could contribute to ill-health or job efficiency effects that may be seen in the working environment.

Air ions are electrically charged forms of the molecules that make up the constituent gases of the atmosphere. There is much controversy over their exact nature, and it is likely that a variety of species and sizes of ions exist and that the physical-chemical properties of negatively charged ions is different from positively charged ones. It is possible that negative ions are charged forms of oxygen and water in the form of $O_2^-(H_2O)_n$ and CO_3^-. Positive ions are more likely to be N_2^+.

These charged particles can be measured by drawing an air sample at a known flow rate between a pair of charged plates. In simple terms, any charged particles in the air will be attracted to the plates and the resulting tiny current can be amplified and displayed. By selecting the sign of the charge on the plate it is possible to measure negative and positive charges separately. A further refinement is to alter the air flow, the charge on the plate, or the length of plates to be able to select and measure different sizes of ions. Much early work on ionization indicated that so called 'small ions', that is, those with a mobility in an electric field of about 1–2 cm $s^{-1}V^{-1}cm^{-1}$ were the only sizes that had a biological influence. Hence commercial ion-analysers are usually pre-set to measure quantitatively charged particles in this mobility range.

If measurements are made of air ions in different situations it is apparent that there are major differences between rural air and city air, between outdoors and indoors, and furthermore that there are a number of factors indoors that further influence the number of free ions in the atmosphere.

Table 1. Typical air ion concentrations found in various locations on fine weather days. Values are averages measured using a 'Medion' air ion analyser which selectively measures small ions in the mobility range 1–2 cm s^{-1}V^{-1}cm^{-1}. Weather, time of day, and fluctuations in pollutant levels would cause large variations in the outdoor values

		Air ions/cc air	
		+ve	−ve
Outdoors	(clean rural air)	1200	1000
Outdoors	(lightly polluted urban air)	800	700
Outdoors	(city air)	500	300
Indoors	(rural; house; no air conditioning)	1000	800
Indoors	(rural; office; air conditioning)	100	100
Indoors	(city; office; air conditioning	150	50

Table 1 shows typical values for negative and positive ions in some of these locations. Whereas outdoors in a rural environment about 1000 ions cm^{-3} of air of each sign would be expected, an air-conditioned office would have in the order of 0–200 ions cm^{-3}. The factors having a major influence on air ion levels indoors are:

(1) Ducted air conditioning. The metal ducting attracts charged particles so that the air is depleted of ions.

(2) Static electricity. Any ions present within a room come under the influence of static charges. In a modern air-conditioned building, especially one with very low humidity, static charges build up on carpets, furniture, wall fabrics, and on the workers' clothing.

(3) Smoke and contamination. The local effect of smoke and dust contamination within an office is just the same as that within a city area and will act as a sponge to mop up ions.

(4) High density of individuals. This has an effect as each individual will remove ions from the air while breathing and each will carry an amount of static electricity. The greater the density of individuals within a room the greater will be the likelihood of air ions being depleted.

Each of these factors in various ways act to reduce (or in some cases entirely deplete) the atmosphere of measureable small ions. The important question to ask is whether or not this ion depletion in any way affects either the efficiency, accuracy or safety with which people do their jobs or the health of people working in an ion depleted environment.

LABORATORY-BASED EXPERIMENTS

We have attempted to answer these questions at two levels. First, we have done experiments in an environmental control room in which temperature, humidity, noise, and lighting can be kept constant, but ion levels can be altered without the knowledge of the subjects or the experimenters. This type of double blind procedure eliminates experimenter bias and any possible psychological effects on the subjects.

Artificial ionization is accomplished by a corona-discharge ionizer. A variety of these are now available commercially but all work on a common principle. If a very high voltage (5–6 kV) is applied to a needle point or fine wire, an ionization corona is formed around the point.

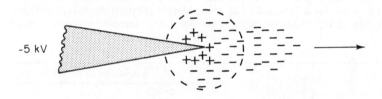

Figure 1. The principle of a corona discharge ionizer

In this corona, pairs of positive and negative ions are formed. If the applied potential is negative, the positive ions are attracted back to the needle and the negative ions are repelled away. These devices therefore provide a constant stream of negative ions, effective within an area of 1 or 2 m from the needle. Some ionizers incorporate an additional fan-blower behind the needle which can propel the ions for a much greater distance. Two such devices would probably be sufficient to produce a very large increase in negative ion levels in a large room. A fan-blower was used in all the experiments. It had an added advantage as one was able to switch off the ionizer at will, while leaving the indication light and blower on all the time. Blind experiments could be set up so that the subjects were always under the impression that they were being treated with ions and it avoided the placebo effects that may have occurred.

In these laboratory-based experiments we find that performance is enhanced in an environment in which there is a predominance of negative ions, compared with low levels of ions or with predominantly positive ions. (Table 2 refers to the ion levels in a typical experiment and Table 3 illustrates some typical results.)

Table 2. Ion concentrations in the environmental chamber during a typical experiment

	Mean ion levels	ions/cc air
Condition	+ve	–ve
'Natural'	250	120
Negative	50	5,700
Positive	5,350	80

Under 'natural' conditions the air was quite depleted of ions because of the metal ducted air-conditioning system. By altering the polarity of the needle we were able to artificially produce either negative or positive ions as in the earliest experiments we wished to differentiate between their effects.

Table 3. Effect of ions on performance. For simplicity performance in the 'natural' ion condition (see Table 2) is called 100. Increments (+) or decrements (–) in performance are percentage changes from the 'natural' baseline level. Statistics are by analysis of variance. n=15 subjects

	% Change in performance		
Task	+ve ions	natural	–ve ions
Mirror drawing	–0.6 ns	100	+28*
Rotary pursuit	+3.4 ns	100	+22*
Visual reaction time	–0.1 ns	100	+6.2*
Auditory reaction time	+2.1 ns	100	+6.0†
Bead threading	–1.5 ns	100	+4.0 ns

ns = not significant at 95%
* = $P<0.001$
† = $P<0.005$

Positive ions appeared not to produce any remarkable effects at all, whether decremental or incremental. However, negative ions did seem to indicate an influence and in all later experiments and field trials we used only negative ions in comparison with natural conditions.

The effect of ions on performance is task-dependent. That is, the effect ions have on performance depends on the task being tested. In general the more complex the task (the more the performance requires complex central nervous system processing), the better the enhancing effect of negative ions. The effect of ions is also subject-dependent. About two-thirds to three-quarters of all individuals appear to be sensitive to ion-changes. In addition we have some evidence that females are more sensitive to ion-depletion and respond more favourably to an ion-enriched environment than do males. The effect of ions on performance appears also to be dependent on temperature and humidity. We do not know the precise relationship between temperature and ion effects but at an ambient temperature of 19°C there is a measurable effect, at 22°C there is a lesser (but still significant effect), whereas at 29°C the effects are not measurable. Thus a relatively low ambient temperature is necessary for artificially generated ions to have a measurable effect on performance. Again, the precise relationships between humidity and ions is not known, but our experiments have shown that at low humidities (30% RH) the effect of ions on performance is much greater than at high humidities (70% RH). This may be caused by ions clustering around the water molecules, as they do around dust particles, making them unavailable to the individual.

In summary, therefore, the effects of negative ions on performance are:

(1) Task-dependent. The more complex the task, the better the effect.
(2) Subject-dependent. Two-thirds to three-quarters of all subjects are 'ion-sensitive'. Females are more responsive than males.
(3) Temperature-dependent. The higher the ambient temperature, the less effective are ions.
(4) Humidity-dependent. The higher the relative humidity, the less effective are ions.

OCCUPATIONAL STUDIES

We have conducted a number of trials to determine whether negative ion generators can influence the health and well-being of people at work. The facts appear to be that a large proportion of people working in modern air conditioned buildings complain of being lethargic (particularly after lunch), of having headache, of feeling nauseous and dizzy, and of having chronic sinus catarrhal infections. The numbers vary considerably from office to office, but often about 50% of a group in an office will have health complaints which they attribute to the office environment. Headache is usually the predominant complaint, with up to about 25% of a group complaining of regular headaches at work.

Although our studies have included an assessment of the effects of room temperature humidity and air ions on comfort and health in a variety of office situations, I will confine the discussion here to a study of ionization in a computer suite, since it is more directly relevant to the theme of this book.

In the study, fifty-four individuals (four females and fifty males) worked a three-shift rota system in a computer-operating area. The shift times were 0830 to 1600 hours, 1600 to 2400 hours, and 2400 to 0830 hours. The part of the suite where the occupants worked for most of the time was equipped with two ion generators (Medion EC300). Each of these generators can effectively treat 9600 ft^3 and whilst operating the room ion levels were about 3500 negative ions cm^{-3}

air and 100 positive ions cm^{-3} air. When not operating, the average ion levels were 500 positive ions cm^{-3} and 550 negative ions cm^{-3}.

Although at this site most of the individuals were computer operators (rather than VDU operators), there were of course VDUs in use and individuals spent varying periods of time at the console. The negative ion levels near to VDUs is rapidly diminished near to the unit because of the high positive charge on the screen. Where many VDUs are operating in a confined space they will undoubtably produce an overall effect of reducing ion levels in the total environment. What is more important is the fact that the micro-environment in which the VDU operator works is far more severely depleted than the room in general. For example, if there are 400 ions cm^{-3} in the general atmosphere this level can diminish to well under 100 ions cm^{-3}, even down to zero, close to the screen.

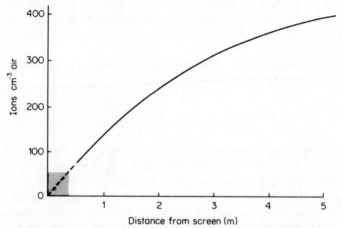

Figure 2. Ion levels near to a 14-in screen with a surface charge of +6 kV. Measurements closer than 0.5 m are inaccurate because of the effect of field strength on the ion analyser (dotted extrapolation). The shaded area is the area in which most VDU operators have their head in relation to the screen

The first four weeks of the trial consisted of the ionizers installed but not working (unknown to the subjects). During weeks 5–12 the ionizers were switched on again unknown to the subjects. The experimenters coding and analyzing the data were not aware of the experimental design, thus ensuring a strictly double blind procedure. At the end of each work period each subject completed a questionnaire (Figure 3). Time of day (shift) has some important influences on subjective rating of environmental hot–cold, pleasant–unpleasant,

Table 4. Summary of the data in relation to ions and shift. The values are means of the subjective rating scores where 0 is the left-hand statement and 100 is the right-hand statement. The statistics (analysis of variance) show the effects of shift, regardless of ions and in addition the effects of ions on each shift. It can be seen that the number of significant effects of ions increases with lateness of shift. P values of less than 0.05 are regarded as significant at greater than 95% probability

| SCALE | IONS OFF WKS 1-4 | | | IONS ON WKS 5-12 | | | STATISTICS SHIFT | | IONS SHIFT 1 | | SHIFT 2 | | SHIFT 3 | |
	1=0830-1600 Mean±SD	2=1600-2400 Mean±SD	3=2400-0830 Mean±SD	1=0830-1600 Mean±SD	2=1600-2400 Mean±SD	3=2400-0830 Mean±SD	F	P	F	P	F	P	F	P
Environment														
Hot–Cold	45.9±13.1	44.9±12.5	50.8±15.0	44.6±13.8	43.3±11.8	47.4±13.0	15.015	<0.001	0.074	0.786	2.794	0.095	5.631	0.018
Pleasant–Unpleasant	57.0±14.5	57.3±12.2	61.2±15.3	54.8±14.9	53.7±14.8	54.8±15.6	3.263	0.039	2.837	0.093	3.824	0.051	17.257	<0.001
Fresh–Stuffy	60.1±15.6	61.6±13.4	62.8±18.7	54.3±17.0	55.2±16.2	55.0±17.9	0.839	0.432	14.224	<0.001	13.363	<0.001	22.860	<0.001
Pleasing–Annoying	57.3±15.2	59.2±12.3	64.0±16.6	54.1±15.7	53.2±15.9	55.3±16.9	7.754	<0.001	4.072	0.044	11.783	<0.001	30.826	<0.001
Comfortable–Uncomfortable	58.6±15.9	59.2±12.2	65.1±16.1	54.9±16.9	53.9±16.8	56.4±17.3	8.510	<0.001	7.113	0.008	7.920	0.005	28.758	<0.001
Good–Bad	58.8±17.3	60.2±12.7	64.0±16.0	54.3±16.5	53.8±17.4	56.0±18.6	5.619	0.004	5.771	0.017	12.588	<0.001	20.782	<0.001
Subjects														
Hot–Cold	44.4±15.8	44.1±14.6	50.3±17.6	43.1±15.0	42.6±13.4	48.1±15.3	20.225	<0.001	0.211	0.646	2.123	0.146	2.121	0.146
Comfortable–Uncomfortable	57.6±14.9	58.4±12.5	64.2±16.1	55.1±15.6	54.4±14.8	56.6±16.6	8.245	<0.001	4.619	0.032	5.553	0.0119	21.791	<0.001
Pleased–Annoyed	55.3±15.8	57.8±14.2	62.7±17.6	51.8±17.5	50.3±16.6	54.2±17.5	12.535	<0.001	2.488	0.115	18.566	<0.001	28.574	<0.001
Alert–Drowsy	50.0±19.5	53.1±19.2	64.4±23.3	42.4±21.0	42.5±22.5	50.4±24.9	32.124	<0.001	8.618	0.003	19.315	<0.001	25.572	<0.001
Well-being:														
Best–Worst	52.2±18.6	54.7±18.3	59.7±20.1	47.4±18.4	45.8±18.8	49.5±19.6	4.566	0.012	0.086	0.771	5.824	0.020	1.197	0.279

176

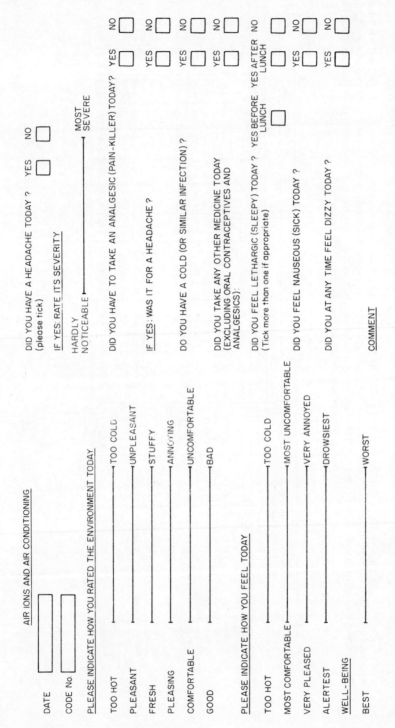

Figure 3. The questionnaire. Subjects are told that statements at either end of the lines (10 cm in the original) should be regarded as the extremes of their experience. They are instructed to make a vertical line to represent their feelings that day between these extremes. It is scored by measuring the distance from the left-hand side of the line to the subject's mark

pleasing–annoying, comfortable–uncomfortable, and good–bad. In addition, subjects themselves showed shift–related changes in hot–cold, comfortable–uncomfortable, pleased–annoyed, alert–drowsy, and well-being scales. In all these responses there was an increased feeling of coldness, discomfort, and displeasure with the later shifts (Table 4). The health questions revealed a similar trend towards an increased number of complaints of headache on the late evening and night shift, but not consistent effects on complaints of dizziness or nausea. The incidence of headache is illustrated in Figure 4.

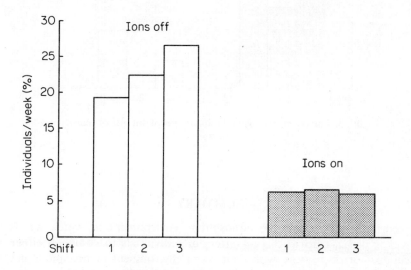

Figure 4. Effects of shift and ions on headache. The incidence is expressed as the percentage of individuals in the group complaining at least once during any week

During the ions-on period there is a very significant improvement in the rating of atmospheric freshness, pleasantness, comfort, and 'good'. In addition, subjects feel more comfortable and alert. The improvements brought about by ionization are even greater on the evening and night shifts, which indicates that ions may have a particularly beneficial effect at night. Ions reduced complaints of headache to a figure of about 6% independent of shift, whereas in the non-ionized period the complaint rate was up to 26.6% and was shift-related (Figure 4).

Ions also had an effect on reducing the complaints of nausea and dizziness (Figure 5).

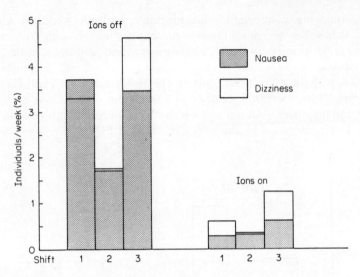

Figure 5. The effect of shift and ions on complaint rate of nausea and
dizziness

SUMMARY

It seems, therefore, that the atmospheric environment in which we work
provokes perhaps up to three-quarters of individuals to complain either of
discomfort or displeasure with their room environment or precipitate actual
complaints of illness. The concentration of small negative ions appears to be one
of the atmospheric factors that is responsible for this. The modern office
environment in which many VDU and computer operators work is usually
deficient in negative ions for a number of reasons, and VDU operators work in a
particularly depleted micro-environment near to the unit because of the static
charge on the screen surface. The introduction of artificially generated negative
ions can significantly improve the subjective rating of the environment and of the
subjects' own feeling of comfort and well-being. In addition, negative ions
significantly reduce the complaint rate of illness, particularly headache, nausea,
and dizziness.

Negative ions also improve certain types of psychological and psychomotor
performance. The generation of negative ions could be a cheap addition to the
VDU console and could bring about major improvements in job efficiency,
health, and comfort. It must be strongly emphasized, however, that ions should
not be considered in isolation of the many atmospheric, environmental,
architectural, circadian, psycho-social, and ergonomic factors that all contribute
in a complex and interrelated way to determine an individual's health at work.

The effectiveness of ions appears to be reduced at high temperatures and high humidities. However, ions have a particularly beneficial effect at night and could contribute in reducing the ill-health and discomfort associated with night working.

Chapter 18

Trade Union Demands: The Changing Pattern

SHEILA MCKECHNIE

Health and Safety Officer, ASTMS

THE EARLY MUDDLES

The trade union movement became aware of the problems of VDUs in two ways. First, there was the growing awareness of the possible effects on jobs of new technology *per se*. The conclusions that all unions came to were that there would be fewer jobs and many of the jobs which remained would be progressively de-skilled, becoming monotonous, boring and totally lacking in job satisfaction. A few million words later that view remains essentially unchanged. There have been many pious hopes, e.g. 'The switch from orthodox handling routines to VDU-based systems provides an opportunity for job enrichment and enlargement. Job flexibility and some control over work allocation should enable the promotion and utilization of individual skills' (Chapter 14, p. 142). I cannot find one instance in the trade union movement in the last five years where this has been achieved.

Second, trade union research departments were beginning to receive requests for information on the health effects of VDUs from members. Many of these queries were vague and often based on anecdotal information. The only specific query which occurred frequently was the issue of radiation output from VDUs.

These two types of pressure on trade unions cannot in practice be separated. Health and safety factors are part of the overall problem of the introduction of new technology. The attempt by many trade unions to limit the time that an operator can spend on a VDU is to ensure that the restructuring of jobs does not create less satisfying work and to minimize any harmful effects on the operator's health, both physical and mental. What has become abundantly clear after five years of negotiating experience on the introduction of new technology is that the collective bargaining system, as we know it in Britain, responds much more readily to the relatively more concrete demands of health and safety standards than the more abstract criteria of job satisfaction. (The reasons for this require a great deal more understanding of the nature of the industrial relations systems than is possible to enlarge on here.) It could also be concluded from these early negotiations that health hazards which could be 'measured' were more easily negotiable than factors which could be described as 'subjective'. I put both factors in inverted commas because the statement which I have made *ad nauseam* to both safety representatives and employers is that because a problem cannot be measured by an agreed scale it does not follow that the problem does not exist. But many safety representatives faced with employers who were contemptuous of reports of operator fatigue, strain, anxiety, etc. started to demand more concrete definitions of the nature of the hazard. The concentration on the issue of radiation in the early trade union documents is an example of this development. The success of employers, if it can really be called success, was to shift the onus of proof on to the trade unions. The fact that most employers and manufacturers had no evidence that the work was safe tended to get lost behind statements that after all it was only a 'television'! It was these underlying factors which explain, and if not entirely excuse, the approach taken by the trade unions when drafting initial guidance to their members. A detailed summary of these recommendations was made by Brian Pearce in Chapter 12 and I see no need to repeat that analysis.

ON-GOING PROBLEMS

It is now clear to trade unions that many of the early problems relating to VDUs were caused by poor quality equipment. It was in response to this that many trade unions attempted to lay down rigid standards which the equipment should meet. Some of these were straightforward and unequivocal, e.g. the keyboard must be detachable. Some were problematic. Recommendations on lighting standards were often a substitute for a poor-quality image on the screen. There was much confusion between the hazards which arose from the equipment itself and those from the interrelation of the environment and the equipment. VDUs were often introduced into office environments that were badly lit, badly ventilated and ergonomically disastrous. These problems were not created by VDUs but the introduction of VDUs brought these problems out.

The nature of the rate of change of phosphor technology itself compounded the problem of laying down standards. The issue of flicker illustrates this. There were many misunderstandings of why the problem of flicker existed, and the negotiated '57 hertz' in one company must be the best example of this. But it was not just misunderstanding that caused problems. A minimum standard had to be laid down in order that safety representatives could be guided on the quality of the equipment. It is not sufficient that trade unions should advise the members to demand a clear, stable image unless they also made available the means to decide what is and what is not a clear, stable image. In the absence of any standard of testing on specific VDU models there was little alternative but to try to lay down minimum standards. This problem will remain unless we can develop a more systematic approach to the design standards of VDUs and, for that matter, all new plant and equipment.

While the trade union approach to equipment standards can be criticized on some points of detail the overall approach was sound. If the problem could be removed at the design stage, this would clearly be preferable to makeshift solutions in the actual workplace. Whether or not this pressure from trade unions was an important factor in the undoubted improvement of design standards on VDUs will, of course, remain speculative but the fact that a number of manufacturers are now marketing their products on the basis of user acceptability has a great deal to do with growing trade union insistence on the need for high standards of design.

While the improvement of standards on equipment will clearly reduce the potential for chronic adverse effects in operators the trades unions are not content to sit back and wait for twenty years to see if, in fact, this is the case. Nor are they willing to accept that only measurable signs of physical disease will be accepted as evidence that VDUs are harmful. This is the whole purpose of monitoring the working environment, even when the risk has been removed, to ensure that it has been effective. Some of the issues and problems over eye tests have arisen because of this. This approach has brought us into conflict with a number of employers and, perhaps more surprisingly, the Employment Medical Advisory Service (EMAS). Two specific issues illustrate the kind of conflict that can arise.

All unions that have issued guidelines have insisted on the operators having regular eye tests. In addition to the documents issued by trade unions discussed in Chapter 12 a number of unions, including ASTMS, have issued specific guidelines on this issue. (*Health and Safety Monitor* Number 9, February 1981 Prevention of Occupational Eyestrain—Hazards of VDUs.) The case against eye tests rests, in our view, on the complacent attitude that there is no definite evidence of harm. APEX, in its comments on the EMAS draft guidance note on the Health Hazards of VDUs, made the following points:

> APEX cannot accept the section referring to Eyesight. The author of the report rightly points out 'very few people have "perfect" vision' but that 'in a great many tasks of a non-demanding nature this is not even a noticeable handicap'. This is corroborated by the authors of the leading work on VDUs, *The VDT Manual*, who state 'between 20% and 30% of the working population have uncorrected or inadequately corrected visual defects'.
>
> Mr Mackay, by implication, admits that working at a VDU terminal is more visually demanding by saying 'There will be some individuals, who on transferring from a visually undemanding job to terminal work experience heightened awareness of visual problems and difficulties in meeting and reading material on the screen or on source documents'. Consequently they

may suffer symptoms of eye fatigue. If this is the case it would appear only common sense to allay fears amongst users of VDUs that VDUs might harm their eyesight by having eyesight tests on employees *before* they use the equipment as well as at subsequent regular intervals.

However, the author of the report attacks these suggestions on the basis that 'there is no evidence from either medical or scientific studies to suggest that VDU operation has a permanent effect upon the eyes or eyesight' and thus 'it would appear unnecessary to specify visual standards for VDU operators which are more stringent than for other clerical workers and, furthermore, it would suggest that regular, routine eyetesting for all operators is unnecessary on medical grounds'.

Firstly evidence from VDU operators would, I am sure, suggest that their work is at least equally, if not more, visually demanding than most other clerical jobs.

Secondly, whilst no medical evidence exists at present, the opportunity has not been taken to build-up evidence, by carrying out the eyetests using the package suggested by the VDU Eye Test (VET) Advisory Group and keeping records of these tests, as they also suggest and hence the 'lack of evidence' that Mr Mackay mentions. Until such records are kept, and only when further research has been completed, can Mr Mackay's statements be accepted or rejected.

The other weakness in this section of the report concerns employees who are spectacle wearers. Contrary to the report, wearers of bifocal spectacles may have problems with VDU work because they are viewed at an intermediary distance (45–70 cm) whilst these spectacles have lenses for near and far-point vision. Problems may also arise with older employees who use reading (near-point) spectacles, as they also may not be able to focus on a VDU screen without discomfort. These difficulties will be identified by the test for intermediary vision recommended by the VET Advisory Group, and can be remedied by the provision, at no cost to the employee, of spectacles incorporating intermediary distance lenses.

I would be less polite. I think the logic of Colin Mackay's argument that eye tests for VDU operators are not necessary because we do not perform such tests for other clerical work can be usefully compared with the argument that because we only have crash barriers on 70% of motorways, we should remove them so everyone can have an equal chance of having an accident!

The second issue that highlights the differences of approach between employers and unions is the problem of work periods. Almost all unions have attempted to negotiate fixed breaks on VDUs and an overall time limit or work on equipment of no more than 4 hours a day. These demands cross over (as was explained earlier) the abstract boundary between physical and psychological effects. They throw into sharp relief fundamental conflicts of interest between trade unions and employers which I do not think are capable of solution without fundamental changes in our society's organization of work. Jobs have not become more satisfying with the introduction of VDUs because of the profound conflict of interests in relation to work design, the interests of the workers, and the interests of the companies buying the equipment. However, retreating to less intractable problems, exposure time to a potential hazard is a well-accepted approach in industrial hygiene to minimizing long-term damage. The reality is that employers have been totally resistant to such attempts to restrict their right to organize work in the way that suits the needs of the organization rather than the needs of the employees. Only in those areas where trade unions have considerable bargaining power has this been achieved.

CHANGES IN DIRECTION

The input of the 'experts' into this discussion has not been particularly helpful and many contributions indicate an appalling ignorance of the authority structures in most workplaces. The best example of this that I have found was as follows: 'In general, the length and frequency of rest pauses depend on a variety of factors, including the task being performed and the working environment. The individual concerned is probably the best judge of when a break is needed' (Purdham, 1980). I do not, of course, disagree with one word of this statement but I do not consider that my chances of incorporating it into a collective agreement would be very great!

The wheel has, in one sense, come full circle. The preoccupation with job design, which was overshadowed by the issue of equipment design, is coming to the fore again. The trades unions are increasingly alive to the fact that when you solve the problems of the equipment and work environment the problem of the nature of the work remains.

The evidence has been growing in many trades unions that the stress-related problems of work on VDUs are much more likely to be the source of detrimental health effects than the design of the equipment itself. Stress is caused primarily by pressure of work, lack of predictability in the response time, confused layout, making the task to be performed more difficult, isolation of the employee, and the interrelation of employee, equipment, and customers in the more common applications of VDUs. The lack of predictability in the response time, especially at peak user periods, causes a great amount of stress. It can take 7, 8 or even 17 seconds for the image to appear on the screen after keying in. The relationship between stress and heart disease is a simple one although the exact mechanisms are controversial. The stress reaction releases chemicals into the blood stream. The very high rate of heart disease in London bus drivers has been attributed to this kind of mechanism. Some unions have already realized the importance of this. At least one has written into an agreement that there will be no bonus schemes or mechanical work-measurements on VDUs. If managements can control bonus schemes on car tracks the need to control bonus schemes based on key depressions is infinitely greater. I would strongly oppose any work measurement on VDUs which is not a group bonus scheme.

In moving into the new area of job design, the trades unions have an even more difficult task than they have with equipment design. If we are to learn from experience perhaps this is an opportunity to analyse how, in my view, the 'system' has failed, or rather has not succeeded in effectively dealing with the problem. By the 'system' I mean the relationship between the responsibilities of designers, employers, the Health and Safety Executive, and the trades unions, which should ensure that the standards of health and safety at work relating to VDUs are agreed and implemented. Academic conferences, however informative, are no substitute for an agreement on standards and effective implementation of these standards.

WHAT WENT WRONG

Here I have looked at all the parties involved and analysed where I feel they have failed in their response to deal with a problem before it had started. I have also tried to look constructively at how they could have picked up these problems with VDUs.

The manufacturers/designers

The design of any piece of equipment takes place against a background of how the designer analyses the priorities in the demands that the design should meet. Such factors might include durability, size, and appearance as well as the requirements of the task the equipment is required to perform. What could be readily shown if we had access to the original specifications is that there is very little attention to user acceptability. This is for two reasons; first, the very low priority given to ergonomic factors in almost all design fields in Britain and second, that we have a second generation of users; computer scientists using VDUs as an appendage to their work are going to respond quite differently to ergonomic problems than those workers with a much lower level of motivation where the VDU becomes central to the job.

For user-acceptability to be an issue at the design stage there would have to be some direct input on ergonomic standards. Where and how is this input to reach the designers? It is too simple to say that it is just good ergonomics and manufacturers do not give ergonomics sufficient priority. This is like saying that all cars should be rust-proofed and why is it not done? The question is how do you make manufacturers meet good ergonomic standards? That is the first issue which an analysis of the history of this problem throws up.

The employers

The criticisms that can be made of employers overlap with those that can be made of designers, but there are many additional aspects for which only employers could have responsibility. Very few employers laid down standards for the buying of equipment. Even where such standards did exist I cannot find one instance where the guidance was specific enough to help the buyer to decide between different models. The ignorance of ergonomics in manufacturing was mirrored by the ignorance of ergonomics in purchasing. Even ICI, to whom I often give credit, has produced a document which can only be called rubbish. It gave no actual advice as to which type of equipment to buy. However, it was only a pamphlet, and perhaps they have another system for giving purchasing advice in that company. In the literature from all the big companies buying VDUs I have found no clear guidance to the management responsible for purchasing. There seems to be no information about what type of equipment should be purchased in relation to its use or in relation to user-acceptability. Consequently I do not think employers have responded effectively.

However, only employers could be responsible for the location of the equipment and the overall effect of it on the working environment. Yet in many offices where ASTMS had members it was literally a 'brown paper and string' effort to adapt the environment so that the VDU could even be used. Heath-Robinson artefacts had nothing on some of the efforts of ASTMS members, and what they lacked in knowledge they certainly made up for it in ingenuity. The conclusion which has to be drawn is that for most employers it was simply an issue of the effectiveness of the system. The question of the working environment and user-acceptability were defined as little problems to get round if the unions insisted.

Many employers under trade union pressure were forced to consult ergonomic consultants after they had decided on the equipment and when the consultant's advice could only ameliorate and not resolve the problem. The question that this poses is, therefore, how do you get employers to introduce new equipment so that the working environment does not deteriorate, or perhaps even more optimistically, to improve it as a result of the introduction of new technology? This does not seem to happen on a voluntary basis.

The law

We have a Health and Safety at Work Act that was primarily introduced to ensure that appropriate standards are set to deal with health and safety hazards. Robens clearly defined the failure of the old system to respond quickly enough to changes in the nature of hazards. The case of the thirteen years it took to agree the Abrasive Wheels Regulations is a common example quoted to make this point. Five years since the introduction of the Act it is still possible to claim that the

Health and Safety Commission did not respond sufficiently quickly to the problem. Nor would the trades unions accept that when a response was forthcoming (the guidance note) that it was an adequate response to the problem. There was already too much 'guidance'; what we needed was enforceable standards. Some of the problems are related to the general absence of specific standards for the office environment. The Offices, Shops and Railway Premises Act was really a piece of arcane nonsense when it was introduced and time has not improved it.

However, if we had had clear standards on lighting, heating, ventilation, etc. perhaps the manufacturers would have been put under more pressure by employers buying new VDUs, as these employers would have the responsibility of making sure that they were compatible with existing environmental standards.

A more detailed analysis of how the Health and Safety Commission should have responded is clearly required. As the system is tripartite, some responsibility clearly rests with the unions themselves. Although many trade union fears about the health and safety aspects of VDUs were conveyed to the HSE there was no attempt to set up any tripartite committee to discuss what should be done. The Health and Safety Executive went about it as if the Act did not exist except, of course, that it consulted the trades unions on the draft guidance note. There was no discussion of what research might be required to set standards, no monitoring to learn from experience, and no forum where it could have been agreed to set up a VDU assessment centre (a suggestion which came out of the discussion in Section 1). The trades unions were faced with the uphill job of simply getting the problem recognized. There may be many reasons for this failure of the HSE to respond, not least the totally disorganized trade union pressure, but any proposals for change will have to recognize and deal with this problem.

The Trades Unions

The trade union response has been analysed in detail by Brian Pearce in Chapter 12. However, as with many problems of this kind, they reveal structural rather than isolated problems. The trades unions in Britain simply do not have the resources to set up their own VDU consultants' team to advise them and to check on what employers are actually doing. To put a trained 'trade union' consultant in with a trade union officer to negotiate the standards in every company in the country is clearly impossible. The only trades unions who ever attempted such an approach were those in the printing industry, but even this could only be a partial approach because of the cost.

The information required to do research is not always available to unions and in many instances we were faced with a situation where our trade union organization was so weak that the equipment was virtually introduced on the employers' terms. If collective bargaining was such a successful method for setting health and safety standards we would not need any health and safety laws at all. This is self-evidently not the case. Trades unions are susceptible to other pressures. The threat to jobs reduces the trade union's bargaining position, to have a job at all becomes the overriding objective in the present economic climate.

Any trade union approach, therefore, must be directed nationally to set the minimum standards and locally to educate the representatives into enforcing these standards. Collective bargaining becomes the method whereby the difficulties of implementation are resolved. It is not, and never will be, an adequate system for setting standards, and in many instances it is powerless to perform even an enforcing function.

Universities

Universities are predominantly state-financed institutions and, while clearly having to be independent of the state in crucial respects, can easily be integrated into the overall system for setting health and safety standards on VDUs. If the HSE did as OSHA (Occupational Health and Safety Administration) does in the United States then that involvement would follow automatically. When OSHA decides that a health or safety issue exists it decides on the work which it requires done to analyse the nature of the hazard and the steps that are required to resolve it. This programme is then listed in the Federal Register and requests are made to tender for research grants. It does not need to be emphasized that adequate resources are a crucial part of this system and this may be the most important key to the whole issue.

WHAT IS TO BE DONE?

The problems can be categorized as follows:
(1)   The control of design standards;
(2)   The control of environmental standards;
(3)   The control of job design factors;
(4)   The monitoring of the workforce.

There are a number of aspects to the solution of all four categories. The standards must be agreed, the responsibility of implementing the standards defined, the enforcement of the standards allocated, and the obligation to monitor the workforce clearly spelled out. The results of this monitoring must be fed back into the system to review existing standards. The following are no more than pointers to the solution of these issues.

Control of Design Standards

A number of voluntary bodies exist providing some kind of advice on design standards. The British Standards Institute could clearly be involved. Other bodies involved in ergonomics, including Loughborough's own HUSAT Research Group are already giving such advice. The purpose of seeking advice would be to allow the HSE in one of its tripartite committees to agree what standards should be met. As the technology is changing so fast this would have to be done in such a way that specific standards could be modified from time to time. Some kind of approval system for new units could be set up providing *Which?* type information on specific units. For this to work successfully such an approval system must either have testing facilities of its own or much stronger powers to have access to employers' information than exists at present. Whether such a unit would be part of the HSE or funding be provided to an outside body would depend on a detailed analysis of the kind of resources required in terms of facilities and personnel.

Control of Environmental Standards

We need new detailed regulations to set standards for office work. Lighting, heating, and ventilation are the priorities. To meet such standards, employers would be required to measure the environment to ensure compliance. Maximum and minimum temperatures are probably the easiest standards to set. How the employer complies with these standards, e.g. by ventilation or air conditioning, etc., is a matter for discussions between employers and unions locally.

Job Design Factors

The only attempt to control the approach which employers take to job design that I am aware of is the Swedish Work Environment Act. This, however, is a major state intervention which is predicated on some very different social structures and values than those which exist at the present time in Britain. The objectives of the Act were as follows:

(1) To secure a working environment which affords the employees full safety against harmful physical and mental influences and which has safety, occupational health, and welfare standards that correspond to the level of technological and social development of the society at large at any time.

(2) To secure sound contract conditions and meaningful occupation for the individual employee.

(3) To provide a basis whereby the enterprises themselves can solve their working environment problems in co-operation with the organizations of employers and employees and under the supervision and guidance of the public authorities.

How the Act works in practice to ensure that jobs do not increase stress or, as the Act says, 'harmful mental influences', it is impossible to say. However, it is rarely possible to take legislation from one social system into another and produce the same effects. But without the same legislative framework I am not optimistic about developments in job design. The interest of human beings will always be secondary to the interests of the business, unless incentives are provided to reverse the position.

Monitoring the Workforce

There are two quite separate needs in respect of this. First, there are studies to be set up to establish whether or not a hazard exists and what are the risks. Some studies have been undertaken in the public and private sectors. The initiative in this respect, however, should have been taken by the HSE. Second, there is the on-going monitoring of the workforce which is needed to ensure that environmental standards are sufficient to ensure a 'no effect' result on the employees. This could be done by simply laying down a standard for the monitoring of employees. Of course, the first problem is to persuade all concerned that this is required. Perhaps employer-resistance is understandable; the kind of resistance that has come from EMAS really is rather more unacceptable to the trades unions.

In conclusion, therefore, I consider that the problem of VDUs highlight the problems in the health and safety system as a whole. Whatever solutions to specific problems came out from conferences and discussions the real failure is that we have no systematic way of dealing with these problems. Again, what we are left with after hours of discussion and millions of words, is just good old British muddle!

REFERENCES

Purdham, J. (1980). *A Review of the Literature on the Health Hazards of Visual Display Units*, Canadian Centre for Occupational Health.

Chapter 19

Humanized Computers:
The Necessity and the Pay-off

T. GILB

Independent Consultant (Norway)

DATA POLLUTION

High-ranking officials publicly admit that the identification numbers which are a by-product of mass data processing systems are a 'burden and nuisance' in the eyes of the public. The Postmaster General of the United States, William Bolger, was reported as saying on 15 September 1978: 'We're not unmindful of the burden and nuisance numbers are looked upon by the public—our social security number, bank account numbers, and everything else.' In the same breath, these executives regretfully announce an added burden, in the name of cost cutting.

Somehow, I feel that we have heard these arguments before: industrial pollution was a necessary evil to provide places of work for people who would otherwise be unemployed.

Now that we have established a sense of priorities regarding the physical pollution of our human environment, it is perhaps time to cast our attention towards the new industrial data pollution; a function of the increase in the new data processing industry.

We are no longer concerned with direct threats to our physical health, but the frustrations of computer systems which are inadequately designed for ease of human use can certainly cause severe psychological distress, when people at the grass roots somehow feel that they are to blame for the errors and delays which often result. And this distress will certainly result in physical ill-health symptoms in some. I would like to stress that I am not merely describing a hypothetical abstraction but a number of cases with which I have personal contact; people whose work situation in connection with computer systems is considered highly frustrating and negative.

193

BELL LABORATORIES FOUND THAT HUMANIZED CODES PAY OFF

As early as 1970 Bell Laboratories in New Jersey conducted extensive research into a narrow but important aspect of the work situation of their many employees in the Bell Corporation (the leading telephone company in the United States). They conducted forty experiments using 350 Bell System personnel (of varying backgrounds). The employees were presented with sixty codes and asked to rate them according to difficulty, where 1 was the most difficult and 9 was the easiest. Each experiment was designed to obtain facts about the ease with which their employees could handle the most common codes on which they had to report many times daily, primarily geographical locations and telephone equipment parts.

The results showed in all respects that the typical abstract numeric codes which were in common use as a result of early data processing traditions, dating back to electromechanical machines before the Second World War (IBM, 1936 and 1956), were a substantial barrier to employee productivity. Meaningful mnemonic codes were considered the easiest to work with, while alphanumeric and numeric codes were found to be more difficult (see Figure 1).

The results showed that if the codes were 'human engineered' in the direction of easy-to-remember codes, which were related to whatever they stood for, then people demonstrated a marked degree of improvement in their ability to learn, remember and reproduce correctly.

Bell's Common Coding Policy

The results were so dramatic, and Bell knew that it could trust the research of its own research laboratories, that it became clear that the corporation could not continue to trust the design abilities of computer specialists, who would continue to design according to their traditions and their personal convenience, without due regard to the much larger questions of large-scale employee productivity, customer service, and accuracy. A twenty-six-point corporate policy was formulated and implemented and each point was backed up by at least one of the research results mentioned earlier (see Appendix). Each policy point was designed to pay off in increased productivity and accuracy, if implemented (Sonntag, 1971; Bell Laboratories Inc., 1970).

Next, teeth were put into the policy by the establishment of a Common Coding Office, which I visited in the summer of 1977 in New Jersey. It was then staffed by sixty people, who served to advise the various US Bell Companies on how to construct human-engineered systems according to the policy. This group acted when systems were going to be changed anyway, they did not insist that everything be changed overnight but they made sure that every change through the years was in the direction of a greater degree of human engineering of the codes used by the personnel.

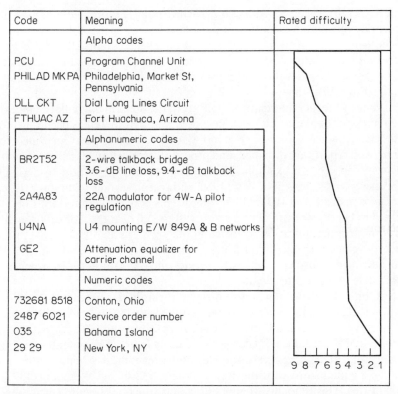

Code	Meaning	Rated difficulty
	Alpha codes	
PCU	Program Channel Unit	
PHILAD MK PA	Philadelphia, Market St, Pennsylvania	
DLL CKT	Dial Long Lines Circuit	
FTHUAC AZ	Fort Huachuca, Arizona	
	Alphanumeric codes	
BR2T52	2-wire talkback bridge 3.6-dB line loss, 9.4-dB talkback loss	
2A4A83	22A modulator for 4W-A pilot regulation	
U4NA	U4 mounting E/W 849A & B networks	
GE2	Attenuation equalizer for carrier channel	
	Numeric codes	
732681 8518	Conton, Ohio	
2487 6021	Service order number	
035	Bahama Island	
29 29	New York, NY	

9 8 7 6 5 4 3 2 1

Figure 1. Preferred codes of Bell Systems personnel (after Sonntag, 1971)

Bell is probably no different from any company or institution where a large amount of employee labour is concerned with data processing and contact with computer systems. People are basically the same, in Bell or elsewhere, in their ability to relate to abstract codes and procedures, or to more natural systems of human communication. If abstract numbers were such a good idea for people, then we should go over to speaking that way. As it is today, such numbers are mainly used as cryptographic codes in order to make it difficult for people to learn the meaning of the message.

I suggest that we are all in need of policies and organizational advisers similar to those of Bell, but we do not all realize this yet. The purpose of this chapter is to increase awareness on the part of levels of influential people as to the nature of this problem and to propose some solutions.

THE COST OF THE HUMAN ELEMENT AS OPPOSED TO THE MACHINE ELEMENT OF WORK

Montgomery Phister has painstakingly collected and massaged several decades of data on the cost of the human and machine element in data processing systems

(Phister, 1976). Using his data as a departure point, I estimated that by 1980 the direct data processing department personnel costs was more than six times that of the computer hardware in the United States (see Figure 2). This is a 300% increase in the relative costs since the childhood of computers thirty to thirty-five years ago. Clearly traditions and priorities developed in bygone years must be changed to reflect this.

In spite of the fact that many people are of the opinion that great changes have been wrought by the computer I have observed that there has been remarkably little change since before the Second World War in the way in which codes, forms, and reports are designed, and procedures for getting data into computers. I have compared, for example, IBM Corporation texts on how to design data processing codes, written in 1936, 1956, and recently, and find that the degree of change is insignificant (IBM, 1936, 1956). This is even clearer when I observe a large number of varied data processing systems all over the world. The methods for human communication to the computers are essentially the same primitive ones developed many decades ago. This fact is only lightly disguised by transferring the obsolete and abstract codes and procedures to electronic on-line interactive computer terminals.

Not only have we failed to consider the rapidly increasing human cost component adequately but we have largely neglected to make use of the available speed and capacity actually available in current computers. We still act as though the speed of a mechanical relay was the limiting factor. The potential for human engineering of data processing codes and procedures to make fuller use of available equipment is documented in detail in Gilb and Weinberg (1977) and Gilb (1976). This is not an attempt to publicize the books but to ensure that one has a specific body of knowledge with which to confront computer specialists when a course of action has been decided. One can expect most of them to deny much of what will be asserted in this chapter because they have been brainwashed differently for so long, and have not been given sufficient motivation to break out of that mould.

The long-term reversal of cost priorities, where it now increasingly pays off to design systems for people, rather than to teach people to adapt to systems, is dramatically illustrated in another prognosis based on Phister's data, which indicate that by 1980 about 70% of the data processing department costs was the human component, as opposed to 40% in 1955 (see Figure 3). We should add to this the important fact that these costs relate only to the very narrow economics of the data processing department. In many companies the true economics of interest in relation to the impact of data processing is in other areas of the company, which can easily be twenty to a hundred times greater, when we remember that the data processing department budget is often about 1–5% of the company budget.

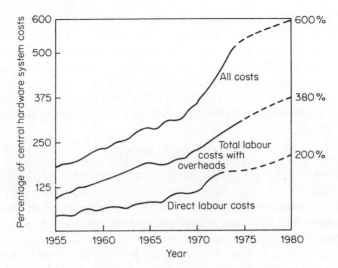

Figure 2. User operating costs (adapted from Phister, 1976).
Hardware cost includes CPU, memory and peripherals, not data
entry, communications, or terminals

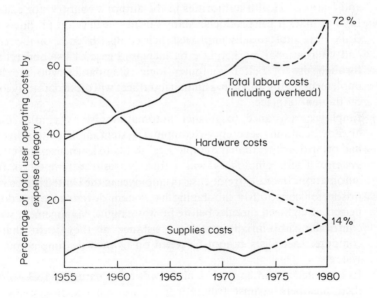

Figure 3. User operating costs (adapted from Phister, 1976)

Still other institutions must consider the economics of the large public which they serve, either for national, ethical or market force economics reasons. These may far outweigh even total internal company or institution economics.

Finally, we should not ignore the fact that in many human engineering situations the critical aspect is security, safety, or utilization of scarce human resources (such as doctors or engineers) and that human engineering of the computer system is often a vital contribution to these aspects as well.

FUTURE THREATS

We are already on the path involving both the misuse of human beings and the under-utilization of opportunity in computer automation. This 'bad architecture' increases the pressure on the institutions which create data-Frankensteins. The following is a list of potential events and trends which will result. Indeed, some, if not most, of them are already happening.

(1)   Loss of business. Customers and the public will turn to other competitors, where they can get more convenient, rapid, and reliable service.

(2)   Loss of service potential. Government and local authorities will be unable to offer the services which are intended. They will have too small a productive capacity to do so as a result of heavy-handed procedures, forms to fill out, delays, and errors. This has already happened to doctors and National Health authorities in the author's country, for example.

(3)   Loss of market for products. Many businesses develop products which could have substantially improved their competitiveness on the national and international marketplace by increased ease of use and reliability through use of better computer logic. Companies who are last to implement highly appealing human interfaces will quite naturally lose out on the marketplace.

(4)   Employee resistance to further automation. Many employees have already been hurt severely by computer system implementations which had priorities other than the employees' ability to learn, use, and effect the system. These employees will naturally resist attempts to further automation. In one extreme case an employee in the United States went to prison for consciously sabotaging his system dozens of times during a period of eighteen months before he was caught. Management will also contribute substantially to this resistance as they learn that their computer specialists cannot be relied on to provide simple and useful systems.

(5)   Exaggerated union control. Unions will quite naturally seek to defend their members against frustrating, unsafe systems or systems which cannot make use of their present skills and background. They have the power and motivation to delay and destroy the profitability of systems which are not tuned to their members' present capabilities and interests.

ENVIRONMENTAL OPPORTUNITY AS A RESULT OF HUMANIZED COMPUTERS

The traditional approach towards computer system development has already alienated the public, the employees, and the unions. Indeed I have frequently seen top and middle management act with varying degrees of scepticism towards further computerization because they equate unreasonably designed computer systems with computer systems in general.

Computer systems, and the abstract codes, detailed forms, unreadable reports, and documents which they produce too often, are clearly a form of environmental pollution. They reduce the quality of life for many of us in annoying and frustrating ways, on a daily basis (Sterling, 1974; Sterling and Laudon, 1976; Tomeski, 1975).

Computer systems which were devoted to helping humans accomplish their tasks with a minimum of distraction and a maximum of politeness, tolerance, and helpfulness would increase the quality of our common environment. We can make them that way if we decide to do so, but senior management, in collaboration with employee representatives, and watched by the press, must create the change. We must deride our early pollution of the data environment, and any continued attempts to continue in that style. We must show that we expect a superior quality of aid from our tool. Computers are the tool we make them. If they pollute our lives, we are getting what we have accepted.

ECONOMIC OPPORTUNITIES FOR HUMAN ENGINEERING OF COMPUTERIZED SYSTEMS

Human engineering may be motivated by security considerations, employee relations, public relations, productivity, or pure good economic investment reasons. This section looks at some of the economic factors which are affected by increased human engineering of computerized systems.

Saving the general public time and effort

This consideration should be made by any large organizations such as government and utility or public transportation which interface with the general public. In some cases such as National Health Insurance (interfacing with medical staff) the same consideration applies.

It is unethical (we could call it social data pollution) to design forms, codes, and reports so that large numbers of users must expend more effort than necessary to handle the data. The cost of design, programming, and computer machine time to save substantial effort on the part of the user public is in most of these cases less than one thousandth of the effort, time, cost, and talent which it is capable of saving.

Since it is unrealistic to expect a computer specialist to consider these 'national' economics, it becomes the responsibility of senior management to ensure that this consideration is given priority, and to audit that it is carried out to the reasonable limit of the technology. The BBC radio and television licence number of about thirty-eight (38!) digits for the general public is a prime example of the ridiculous lengths technologists will go to when management does not pull the strings properly. The burden many medical people have in filling out forms for other authorities is another example of unacceptable misuse of scarce and critical human effort, which can be stemmed only when management and the profession affected decide to stop these outrages. They are only tolerated today because too few people realize that they are to a large degree unnecessary with present and future economics and technology.

Saving employees' time and effort

A fundamental application of human engineering (I mean using all the tricks of the human engineering trade) in certain areas of data processing can increase human productivity by a substantial amount. As a rule of thumb, in redesigning a data processing application using the humanized input principles we can always expect an easily measurable reduction in work operations of a factor of three to ten (Gilb and Weinberg, 1977; Gilb, 1978).

Hedge against inflation and variable workforce supply

Increased human engineering is the same thing as increased automation. We make a current investment in design, programming, and computer machinery in order to save human effort of a repetitive nature for many years to come. An increased part of our workforce is automated, and can be 'hired and fired' without human consequences, on an as-needed basis. This avoids human, union, social, and legal problems of hiring optimistically during boom times and later having to shed it (or worse, keep it on until reserve capital is drained to the point of threatening the livelihood of the core of employees).

Increased market share due to more competitive services or products

Fundamental human engineering of computerized systems is so rare at the moment that any company which takes the opportunity to apply it will naturally become a leader in this area within their industry or service. Human convenience is known to be a major factor in many areas for determining buying or products and continued use of repetitive services. There is every reason to expect a successful assault on this area to result in an increased market share, although this applies only to the leaders, since in the long run everybody must join in to compete. I wish that my banks and insurance companies, for example, understood this to a far greater degree than they seem to do. They are moving in that direction, clearly, but they have missed many present opportunities for many years because of a lack of management guidance in this area.

TECHNICAL OPPORTUNITIES FOR HUMAN ENGINEERING OF COMPUTERIZED SYSTEMS

The following list is a partial sample of specific technical possibilities which can realistically be considered by most non-trivial computer system designs in order to improve the human productivity related to the system. In many cases these objectives can be accomplished by suitable additions to current systems. Supporting technical detail is available in the references, particularly Gilb and Weinberg (1977).

(1) Reduction or elimination of forms to fill out. In many cases the computer input form is no longer technically necessary. It is a relic of earlier and more limited data processing technologies or economics, and often represents an extra work step, with delays, additional work, and additional error possibilities. In most cases a far more direct approach is possible between the source of the data and the computer. Most computer professionals are not as yet trained to handle this design problem, but there is nothing particularly difficult about doing it, once they are given the motivation to accomplish improvements here.

(2) Simplification of codes presently used in connection with computers. Most of the currently favoured long numeric codes, like the ones on a credit card or bank account, are primarily a result of technological limitations which are not, or will not be, valid in the future. These codes primarily serve narrow machine-to-machine communication needs, and might perhaps have some limited use within machines in the future. But at the interface between humans and machines we can use the codes people naturally prefer such as names and addresses, even when these are varied in content and correctness. The job of finding the correct record in computer files can be left to the machine, or to a dialogue between people and machines.

(3) Processing 'partial' information and doing so in 'varied' sequences. Due to technical limitations of early data processing technology there arose a strong tradition of insisting on a complete set of information ('filling out the entire form'), and making sure it always came in the system in a predetermined sequence. This is no longer necessary, since the logical ability of computers coupled with the ability to retrieve or deduce missing information based on computer database information, allows humans to communicate in the natural sequence which the outside environment often thrusts upon us, and with the natural incompletenesses which occur (often because outsiders correctly believe we already have that

information stored in our files). The computer is fully capable of determining what information it really needs to solve present tasks, and will help us to keep our information gathering to a necessary minimum.

(4) Tailoring computer output page content to the recipient. Again, based on early computer limitations in storage for logic and texts, we have accepted as routine a high degree of standardization of forms printed out by computers. Too much information is often pre-printed, to save the computer doing that job. This can confuse the reader. The reader is expected to interpret codes (such as '65' on an invoice meaning Value Added Tax) hopefully printed on the back of the form, but which could better be spelled out directly. Small business computers have many of the limitations of early computers from the 1960s and this increases the temptation to continue with these traditions, although even the smallest computers of the 1980s are capable of a much higher degree of human engineering than is presently being practised by most systems of any size.

The opportunity is there to design printed matter with an extremely high degree of orientation to the parameters of the user (type of customer, geographical district, previous buying history, special requests for certain types of data) (Feeney and Hood, 1977). This in turn will do much to ensure a smooth, error-free connection with the recipient. The necessary investment in more highly tailored programs should not exceed 1–5% of the project cost of not doing this tailoring.

(5) Letting the computer observe 'normal behaviour' and adapt to it. Before computers, humans quite naturally observed and recorded certain patterns of the environment which they were serving. They could not avoid doing so. In most present computer programs this power of observation is not programmed, or if the data is collected, the system is not programmed to make use of it.

As an example of simple things which can be done, if a customer or user has normally requested some special service, such as a few extra copies of an invoice, and an order comes from that customer without explicitly asking for that service, the computer could have been programmed to give it to him anyway on the grounds that he probably wants it. A message could even indicate that the computer has guessed that he wants it, and if he does not he can say so next time. Another example would be keeping track of normal patterns of amounts of money recorded in a financial accounting system, both with regard to the highest amounts normally deposited or recorded in particular accounts, and even the nature of the amount (even numbers, numbers divisible by 5, last two digits 95, 98, or 99). These patterns can be recorded individually and can be used to check future amounts and to give 'suspicious' warnings to clerks, professionals, or auditors depending on needs.

THE SYSTEM USER'S BILL OF RIGHTS

Both employees and outsiders affected by a system, the general public and employees of other institutions and companies, should have certain rights to expect a reasonable standard of human engineering in computer systems. The humanization of the system is normally of benefit to the system-creator institution, so the main function of the bill of rights is to remind the system architects of the duty they are expected to perform, and to give them support in humanizing the systems. The form of this 'bill of rights' is as follows:

(1) System forms, instructions, and outputs should be easily, immediately, and reliably understandable to all people who must make use of the system.

(2) When the users of a computerized system, or any extension of it such as a form to fill out or codes to write down, are in doubt as to the correct actions, they should have access to immediate advice, if possible from the computer itself.

(3) Computer systems should be programmed to understand most of the commonly used formats of data. It should not be a requirement that people learn the exact arbitrary method which the computer programmer finds convenient. The computer should be required to learn most forms that people commonly find convenient for expressing themselves.

(4) The sequence of giving data or instructions to the computer should be capable of being varied to suit the environment of the user of the system.

(5) The system should be adapted to people and their environment, habits, traditions, and their tendency to human error.

(6) Codes should be those preferred by their users, since these will be the most reliable and efficient in terms of total system economics.

(7) Systems should be designed for successful use with the least-skilled people within the expected group of users.

(8) When users show signs of error, non-co-operation, dissatisfaction, or lack of productivity in using a system, all parties should put the responsibility for this failure on the design of the system, never the users; and expect satisfactory changes to be made in the design of the system.

(9) Wherever and whenever the machine can take a burden from human minds and shoulders, we have a right to expect this to be seriously considered.

(10) The user is always right.

MANAGEMENT GUIDELINES FOR HUMANIZED COMPUTER SYSTEMS

Here are some possible courses of action for consideration.

(1) Commission a study of the potential for increased automation degree in present computerized systems. Who is affected, by how much, and what would changes cost?

(2) Develop a company policy to increase the degree of automated help to employees and the public served, to the furthest extent presently practicable.

(3) Initiate a short-term small-scale change in a present system in order to get a feel for the potential offered. Record the productivity and reliability increase, and employee or public opinion regarding this change. This is a pilot study to give people experience and confidence.

(4) Ask data processing management to ensure that all computer professionals are reasonably conversant with the new humanized technical design methods, and that at least one is thoroughly trained as a specialist.

(5) Involve employee representatives. Get them to voice the changes which will make their work environment more productive, reliable, and comfortable.

(6) Insist that all new projects have specific measurable goals regarding the human productivity factors in the data processing systems. These include work production, error-freedom, degree of training, and personal opinion of satisfactory work environment. The management and employee representatives who will effect the system design must be trained in the method for expressing their needs in this area.

(7) Make sure that all major and long-term projects have planned both automated and organizational feedback mechanisms so that signals for necessary change are captured and acted upon. The computerized system must be continually evolving and adapting to current real world events.

(8) Survey external and public aspects of present systems. How do customers and members of the general public react to computer outputs and to the forms they have to fill out? Identify at least one area which could be substantially improved by means of a new 'interface' to the old system. Make at least one such change and measure the effects of the change.

(9) Continue to give signals and questions to managers to impress upon them the priority of making the best use of human and computer resources. With a view towards the long-term future of systems being developed now, encourage them to overdo humanization, rather than take a chance on missing these opportunities.

(10) If the experts you now listen to cannot speak your language, get one who can.

APPENDIX: CODE DESIGN PRINCIPLES

General

(1) Codes should be designed for the least skilled workers within the expected population of code users.

(2) The preference of code users should be taken into consideration.

(3) The total set of code systems in use by intended users of a new system should be taken into consideration.

(4) Each code should be unique.

(5) In defining size of vocabulary from which the code system is to be drawn, room must be allowed for anticipated growth of the code system.

(6) Numeric characters should be given preferences in codes designed for simple tasks.

(7) Since it is easier to convey meaning with alpha characters than with numerics, tasks requiring complex internal processing of the code should use alpha to the extent possible.

(8) Meaningfulness should be built into codes by whatever means and in as many ways as possible.

(9) Where meaning cannot be equally provided for all codes within the code system, preference should be given to those codes which will have the highest use frequency.

(10) The length of codes for a given system should be optimized in terms of the code users' capabilities.

Code length

(11) The code designer should take advantage of common English usage practices in physically dividing or connecting long code phrases.

(12) Techniques for psychologically shortening long codes or code phrases should be capitalized upon wherever possible.

(13) Trade-offs in costs in artificially lengthening codes and code phrases should be carefully considered.

(14) Codes should be made up of characters already in the workers' repertoire.

(15) The vocabulary from which a code system is drawn should be kept as small as possible consistent with unique coding and anticipated growth.

(16) The vocabulary for a given code system should contain the fewest possible character classes.

(17) Rules for the design of codes for any given system should be clearly defined and consistently followed.

(18) Symbols should be used consistently for the same meaning or function.

(19) Where size of vocabulary permits, characters which are likely to be confused visually or acoustically should be avoided.

(20) Special techniques to aid in discrimination of perceptually similar characters should be consistently employed.
(21) Where options are available, simple characters should be given preference over complex characters.
(22) Where options are available, the effects of the serial position of characters should be used to enhance performance.

Formatting

(23) Code elements and phrases should be formatted in terms of the users' order of needs for information.
(24) Decisions on formatting messages or on arrangements of codes within a phrase should be co-ordinated among system users.
(25) The length of the coded message should be kept to the minimum information required by prospective users.
(26) Formatting of coded messages should be such as to facilitate ease of scanning for accuracy and completeness.

REFERENCES

Bell Laboratories Inc. (1970). 'Common language coding guide', Bell Laboratories internal publication.

Feeney, W.R., and Hood, J. (1977). 'Adaptive man/computer interfaces: information systems which take account of the user', in *Computer Personnel SIGCPR of ACM*, Vol. 6, Nos. 3–4, San Diego State University, 4–11.

Gilb, T. (1976). *Data Engineering*, Studentlitteratur, Lund, Sweden.

Gilb, T. (1978). 'New generation of input technology', in *Data Processing (UK)*, Part I (June 1978) and Part II (July/August 1978).

Gilb, T., and Weinberg, G.M. (1977). *Humanised Input: Techniques for Reliable Keyed Input*, Winthrop Publishers Inc., Cambridge, Ma.

IBM (1936). *Preparation and Use of Codes*, Handbook code AM-5, Series Machine Methods of Accounting.

IBM (1956). 'Modern coding methods', IBM form 32-3793-6.

Phister, M. (1976). *Data Processing Technology and Economics*, Santa Monica Publishing Co., Santa Monica, Ca.

Sonntag, L. (1971). 'Designing human-oriented codes', *Bell Laboratories Record*, 43–9.

Sterling, T.D. (1974). Guidelines for Humanising Computerised Information Systems CACM, 609–13.

Sterling, T.D., and Laudon, K. (1976). 'Humanising information systems', *Data-mation*, 53–9.

Tomeski, E.A. (1975). 'Building human factors into computer applications', *Management Datametrics*, IFIP-Adm Gr, Amsterdam, Vol. 4 No. 4, 115–20.

Chapter 20

Measures of User Acceptability

LEELA DAMODARAN

Loughborough University of Technology, Loughborough, UK

INTRODUCTION

Three issues have emerged during the earlier chapters which are of special interest to me. First, we are all too familiar with gross generalizations about human behaviour, for example, that users are resistant to change or have certain stereotypic responses to new technology. Yet all of us who work in this field are very well aware that human reactions to technology are varied and complex and that there is a need for systematic investigation and measurement of user acceptability and attitudes. Second, the need for co-ordination between ergonomics and industry has also been made apparent in this book. Sheila McKechnie made an important point by indicating specific ways in which various contributions can be drawn together to avoid the absurdities of decision-making which may occur when disparate bodies give contradictory guidance. We lose a great deal by failing to share knowledge and responsibility. Third, the issue of failure to utilize available technology has emerged. There is a vast technology concerned with planned change which is generally ignored in the introduction of new computer technology. A great deal of knowledge exists that is now concerned with humanizing many aspects of computer systems but there is a widespread failure to apply it effectively.

All three issues are concerned with problems of acquisition and utilization of human factors expertise. Perhaps our next conference should address itself to ways of overcoming the obstacles and barriers to effective use of technology.

THE NEED TO MEASURE ACCEPTABILITY

Concern for the physical health of VDU operators has led in recent years to consideration of potential hazards such as cataract and facial rashes. Research into the impact of VDU usage upon the health of VDU operators has focused upon these serious medical problems and has sought to apply clinical criteria to establish the existence and the extent of likely hazards.

Research has also indicated other ways in which the use of VDUs can lead to a deterioration in user well-being. Visual and postural fatigue, headaches, boredom, frustration, and loss of job interest and satisfaction are frequently reported by VDU users. These effects may be less dramatic than the serious medical problems, but they can be psychologically serious and important for the operator and will affect task performance. Many of these ill-effects appear to result from mismatches between the physical environment, the hardware and software on the one hand, and human characteristics on the other. In many cases these problems, although persistent, fail to be rectified because only the users are aware of them. Despite difficulties, most users continue to operate the VDU system and therefore, in the absence of routinely applied assessment procedures, the problems remain hidden. One frequently hears the classic quote 'the operators always make that mistake' from managers and supervisors of VDU users. Investigation quickly reveals that the designers of the systems have contravened perhaps every aspect of Tom Gilb's bill of rights for the user. The quoted comment accepts and recognizes that persistent difficulties exist which are not being investigated, and reveals the mismatch between the design and the human characteristics.

A small minority of users may cease to use a system which they find unsatisfactory but such users will usually be high-status senior members of the organization who have discretion over the 'tools' they utilize in their work. Typically their rejection of the system is explained away in terms of personality characteristics, or excessive work load and job pressures. For the great majority of VDU users, however, there is no alternative but to operate the VDU despite adverse experiences. The manifestations of ill-effects of VDU usage upon these users are likely to be revealed indirectly, for example, in high error rates, loss of productivity, absenteeism, high labour turnover, increased sickness, etc.

There are important economic reasons for investigating the problems; it is not solely a humanitarian concern. Industry pays a heavy price for the kinds of ill-effects that result. These may not always be as emotive or dramatic as the risks of cataracts, sterility, or damage to unborn babies and the like but may have a profound and persistent impact upon the performance of our technological and human systems. The ill-effects are frequently persistent and yet are unlikely to come to the notice of those with the authority to rectify them. Part of the reason for the problems remaining hidden is because of the sheer fact of human adaptability.

To gain a realistic appraisal of the impact of VDU usage upon the well-being of the user therefore requires systems to be evaluated on a regular basis, not from a technical standpoint but from the user's perpective. The first task in rectifying these problems is to detect the existence of any problems and their magnitude. This chapter suggests that systems should incorporate regular audits of system acceptability in order to feed back to managers and systems staff the degree to which the system is meeting user needs. The HUSAT Research Group has been refining a set of measures for system evaluation from the user's perspective.

Generally there are no regular appraisals of user acceptability. This is a strange phenomenon when one realizes that most systems are regularly appraised for their technical and economic performance. Any evidence which does exist concerning user acceptability is frequently obtained by rather indirect methods. System evaluations can identify a degree of misuse, disuse, or underuse of a system and very rarely do we find computer systems being exploited as effectively and efficiently as had been intended by the designers or purchasers of that system.

Error rates and complaints of ill-effects are very useful indirect indicators to the problems of acceptability to the user but do not tell us how to rectify the problems that they indicate. The starting point for developing relevant measures of acceptability must be the investigation of the user's experience and perception of the system. This may seem obvious, but all too often concepts of user friendliness incorporated into a system are those of the system's analysts and designers, not the users. Far too many presumptions are made about what the user needs and will find acceptable and easy to use. The real requirement is for investigation and thorough analysis of what the user actually requires.

THE FUNCTION OF SYSTEM ACCEPTABILITY MEASURES

In the system design process there is a need for techniques which permit assessment of likely system acceptability both before and after implementation. As a predictive aid to the designer, system acceptability measures offer a technique for identifying likely problem areas in such a way that remedial action can be planned. Later the techniques offer a means of gaining subjective assessments of the system based on users' experience of the system in operation in order that relevant modifications can be made. It is important to emphasize that the techniques should be used to indicate in a systematic way the problem areas to be tackled rather than simply gaining general expressions of positive or negative attitudes towards the system. Thus, there are two major applications of 'user acceptability measures':
   (1)   To establish user needs during the design process; i.e. what would make a system acceptable?
   (2)   To assess, after implementation, the subjective extent to acceptability.

USER ACCEPTABILITY INDICES

The design of the environment, hardware, and software and the quality of user support all play a part in determining the acceptability of a VDU-based system. Relevance and appropriateness of the computer application for the user's job itself are important criteria upon which users assess systems from the moment the system becomes 'live'. In the long-term effects of the computer system upon job satisfaction, work load, career prospects, and so on become crucial considerations in system acceptability. Thus it is clear that the user's evaluation of a system is influenced in diverse ways by all his experiences, direct and indirect, of the system.

Empirical research (Eason *et al.*, 1974) has revealed that users typically evaluate systems on the basis of characteristics which can be grouped to provide four indices of user acceptability:

(1) Task Fit;
(2) Ease of Use;   } Early, immediate impact upon the user;
(3) User support;
(4) Indirect Consequences: Long-term effects upon the user.

Each index is derived by gaining responses from users to operations relating to a number of constituent items. Examination of the items which comprise the four indices will reveal that the indices are by no means mutually exclusive or independent. For example, 'extra work' which arises for the user as a result of a poor match between his task needs and the characteristics of the output he receives is an item in the Task Fit as well as in Ease of Use assessments.

Task Fit Index

This index measures the extent to which the system characteristics are matched to the nature of the work which the user performs (i.e. fundamental user task characteristics). The user's task can be defined at many levels. The Task Fit Index reflects the user's preoccupation with performing the primary task or job function which the computer system was intended to support. For example, in assessing system acceptability a stock controller would attach great importance to the system's capability to improve his ability to control stock movements. Similarly, a manager's assessment of a computer system with financial modelling facilities will be based upon the degree to which it helps him in his forward planning and related tasks. Thus some of the attributes upon which users assess the appropriateness of the computer aid provided are indicated by empirical research to be:

Accuracy;
Relevance;
Recency (i.e. is it up to date?);
Timeliness (i.e. does it arrive at the right time?);
Completeness;
Extra work (i.e. does it require transforming into a different format, other units, etc.?);
Difficulties created by breakdowns/systems failures.

Thus the Task Fit Index measures the degree to which the service from the system matches the task needs experienced by the user. A similar set of factors was derived from a factor analysis of 'user perceived quality' of computer service by Dzida *et al.* (1978).

Ease of Use Index

This index provides a measure of the effort, difficulty, or strain involved in using the system. This is in some ways the best developed set of items. The emphasis upon visual display ergonomics over many years has given rise to detailed human factors specifications to provide ease of use. The focus has been largely upon hardware, e.g. keyboard characteristics, screen angle, and tilt etc. Major aspects of ease of use relating to software have received rather inadequate attention although the situation is changing slowly. Current interest in 'user-friendly' characteristics has highlighted the importance of acceptable software. Unfortunately, little effort is made to identify the user's real needs in most cases. The point has already been made that all too often 'user-friendly' procedures are designed on the basis of what systems personnel believe the user will find acceptable rather than upon systematic investigation of the user's actual needs and preferences. Appropriate use of system acceptability measures—particularly the Ease of Use Index—could provide a relevant guide to designing systems which really are 'user-friendly'.

Items about which users characteristically express concern in the context of acceptability and ease of use include:

'Terminal ergonomics'—characteristics of work-station, screen, and keyboard;
Difficulties with operating procedures;
Extra work;
Response time;
Characteristics of output—e.g. legibility, format, etc.

User-Support Index

Effectiveness of user support is rarely recognized as a major determinant of system acceptability yet the evidence of its impact is steadily growing. User support is generally approached in a fragmented way. For example, different organizations and different individuals focus upon specific aspects of support such as within-system aids, 'Help' facilities, or initial training, but there is no co-ordinated design of a user support network. This is desirable, as support needs vary considerably with the experience of the user. The extent to which users and system performance suffer as a result of this lack of co-ordinated and effective support is revealed by investigations of existing support mechanisms. In assessing acceptability of a system, users frequently place considerable emphasis upon the availability and quality of help, advice, and instruction. Thus the User-Support Index comprises items concerned with the availability and quality of

Training;
Manuals, guides;
Within-system aids, e.g. 'Help' facilities;
Human support/advice;
Procedures for handling faults/breakdowns.

The real requirement is for a network of human and documentary support facilities. Its effectiveness can be assessed by questioning the availability and quality of training, manuals, guides, within-system aids, and human support. In particular it is necessary to determine whether the help is available when and where needed and if in the form of support that the user requires.

Indirect Consequences Index

The 'index' of acceptability of indirect consequences that has been developed comprises a range of organizational and job-related aspects which are affected by computer systems. The attributes can usefully be considered in two groups:
  (1) Organizational factors, e.g. methods of work organization, policy on new technology, management style, supervisory practices, information availability, etc.
  (2) Job design factors, e.g. skill levels, degree of pacing, autonomy, variety, career development opportunities, work load, communication opportunities.

The effects of the system on important aspects of people's lives is one consideration often overlooked. When investigating the acceptability of a system we often fail to recognize that a computer can have dramatic effects on job satisfaction, work load, career prospects, communication patterns, and many things which are one step removed from the immediate computer interface. Many of these issues profoundly affect the way people perceive systems and the extent to which they regard them as acceptable or otherwise.

Investigations of indirect consequences of computer systems (in collaborative work in Europe) (Eason et al., 1977) have led to considerable elaboration and refinement of these measures and the related analytic techniques. There are therefore considerable data available now to develop guidelines for job design and work organization in automated systems. A great many of the techniques, procedures, and mechanisms are known and available for achieving desirable indirect consequences, yet there seems to be widespread failure to exploit those capabilities. There are exceptions and a number of organizations are committed to improving job design and to finding ways of ensuring that automated systems do match the demands of human beings.

APPLICATION OF ACCEPTABILITY MEASURES

User assessments of the acceptability of various aspects of the VDU-based system are gained through a semi-structured interview. The interview is structured to lead the user through his experiences with the computer system. Each item of the various indices is explored with the user. On the basis of the users' responses a simple numerical score is calculated to provide measures of Task Fit, Ease of Use, User Support, and Indirect Consequences.

At least as important as the numerical score, however, are the qualitative data which result from asking questions as a follow-up to users' responses on the various items probed in the interview. The follow-up questions examine the basis for the users' evaluations. Thus if a user indicates, for example, that the VDU is not easy to operate, then clarification is sought to establish the nature of the difficulties. Very often, when the users can identify a problem they are already well on the way to generating a solution because they have lived with that problem for so long. These data are clearly essential if the objective of the audit is to establish the shortcomings experienced by the user with a view to overcoming them. It is important that the interview or survey technique is sufficiently unstructured to permit the user to reveal his own perception of the system. The semi-structured nature of the interview demands a high level of interviewing skill on the part of the interviewer and an understanding of the VDU operation. Without some understanding of the purpose, function, and operation of the system it is likely to prove impossible for the interviewer to formulate appropriate follow-up questions.

It is crucial that any measures of acceptability are not applied in a mechanistic way so that they simply turn into self-completion questionnaires with a degradation of responses as users lose interest in completing them. In a recent book (Coombs and Alty, 1981) there are several papers which discuss the role of interviewing as part of a counselling, evaluating, and user support role in the provision of computing systems within universities. Many of the lessons are relevant to other settings.

User acceptability measures need to be applied with considerable care if they are to be of value. Decisions on the timing and frequency of user interviews and the choice of interviewers require careful consideration. The issue of who is appropriate to conduct user interviews needs to take into account a number of factors, such as the acceptability of the individual to the users, his/her interviewing skills, and his/her role and position in the organization. The interviewer must be respected, trusted, and regarded as reliable, and must have integrity if the users are to be frank about their experience of the system. It is important that the interviewer can give reassurances about anonymity and provide a certain degree of protection to the user by having some control over the use of the data he collects. In many organizations it is possible to identify individuals who are already accepted by users and who are influential in determining the effectiveness of the VDU-based system. These individuals frequently fulfil some kind of user-support role. For example, knowledgeable users who provide help and advice to fellow-users frequently become informed 'local experts'. These individuals or, perhaps, user liaison officers, could conduct the user interviews as part of their jobs. Their job responsibilities could then incorporate regular audits of user acceptability and remedial action in response to user comments and suggestions. Alternatively, independent interviewers from outside the organization could be used where impartiality is considered particularly important or where there is a poor relationship between the systems personnel and the end users. The essential requirement is for regular, systematic appraisals of system acceptability to be conducted by a trusted and competent interviewer.

Organizational climate and the prevailing management style will also have a significant influence over the way in which user-acceptability measures are administered and the findings utilized. Before embarking upon an audit of a user's evaluation of a system it is important to consider the full implications of such a step. Interviews concerned with assessing user acceptability are as open to abuse as are any other interview surveys. In other words, findings can be used for purposes quite disparate from those the interviews were intended to serve. Thus for example, the data processing manager who wants to eliminate certain practices may see a user-interview survey as a opportunity to identify the 'offenders'. Any such misuse of interview findings can ensure that the acceptability measures rapidly become valueless. Users will quickly learn to distrust the objectives of the interviews and are likely to withhold or distort the information required from them. To be successful in gaining measures of acceptability it is important to make clear to users that findings are to be used to improve the system as they experience it, and not for other purposes. Only by ensuring that users are not penalized in any way for giving their views, however inconvenient or unsatisfactory these are from the point of view of managers or computer personnel, can co-operation be expected. Positive action to remedy shortcomings revealed by the users will also be required if user motivation is to be maintained.

One of the major dangers of conducting an attitude survey is that you raise expectations that things will be changed or improved. If it is concluded that changes would prove too expensive or that the operators are unreasonable so that no alterations are made, then the original situation may be worsened, leading to loss of morale and motivation.

EXPERIENCE WITH SYSTEM-ACCEPTABILITY MEASURES

The four indices outlined above have been applied usefully in varied applications for two distinct purposes. The more common function of the indices has been to establish the degree to which computer applications are acceptable to their users and to indicate problem areas requiring remedial action. In addition, the indices have been used in a design situation as a basis upon which to generate human factor design criteria. In other words, it has proved useful in the system design process to set objectives with regard to Task Fit, Ease of Use, etc. This has begun to apear in new systems design texts, for example, in the PORGI handbook (Kolf, 1979) and in the manual we have prepared with the NCC (Damodaran *et al.*, 1980).

This should in some measure meet the requirements called for earlier. Some of the human factors data that does exist should be put into operation ahead of system design and not wait until problems occur which indicate the shortcomings of the hardware or software. The criteria should be established at the outset to ensure that human factors designs are used in conjunction with economic, business and technical criteria in the selection of systems.

THE FUTURE

Only by monitoring acceptability and continuously improving and maintaining it can user well-being be promoted. The time for a systematic appraisal of acceptability is overdue. It is clear from research data that good user support is an essential element in user acceptability, and one way to ensure that effective audits are conducted is by building them into user-support networks. On the basis of considerable experience gained through research and consultancy it is believed that system-acceptability measures offer considerable scope not just for modifying existing systems but to promote design of new systems which better meet human characteristics.

REFERENCES

Coombs, M., and Alty, J. (eds) (1981). *Computer Skills and the User Interface*, Academic Press, London.

Damodaran, L., Simpson, A., and Wilson, P.A. (1980). *Designing Systems for People*, NCC, Manchester.

Dzida, W., Herda, S., and Itzfeldt, W.D. (1978). 'User-perceived quality of interactive systems', *IEEE Transactions on Software Engineering*, **SE-4**, No. 4, July, 270-5.

Eason, K.D., Damodaran, L., and Stewart, T.F.M. (1974). 'A survey of man-computer interaction in commercial applications', *LUTERG Report No. 144*, Department of Human Sciences, Loughborough University of Technology.

Eason, K.D., Stewart, T.F.M., and Damodaran, L. (1977). 'Case studies in the impact of computer based information systems upon management' *Report No. DHS 243*, Department of Human Sciences, Loughborough University of Technology.

Kolf, F. (1980). 'Guidelines for the organisational implementation of information systems—concepts and experiences with the PORGI implementation handbook', in *The Human Side of Information Processing*, ed. by N.B. Andersen, North-Holland, Amsterdam.

Chapter 21

Job Design and VDU Operation

K. D. EASON

Department of Human Sciences,
Loughborough University of Technology, Loughborough, UK

THE OCCURRENCE OF HEALTH HAZARDS

The debate about the potential health hazards of VDU operation is largely related to a particular kind of job in which an operator sits at a VDU workstation for long periods and engages in relatively repetitive operations with the VDU. As Figure 1 illustrates, this is a fairly small sub-set of the total population of VDU users.

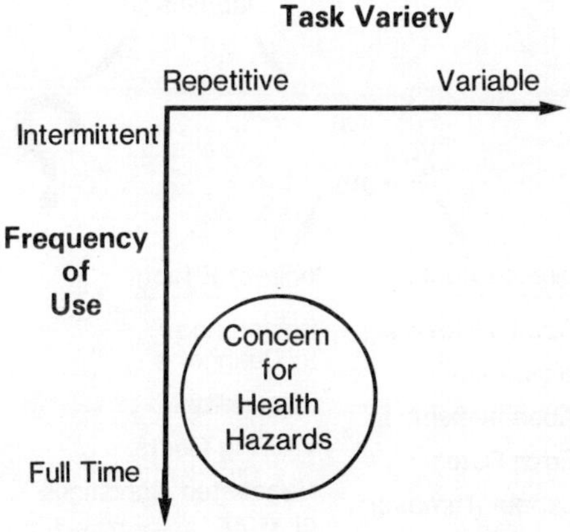

Figure 1. VDU operation and concern for health hazards

It is rare to find reports about intermittent users of VDUs, or users who have complex and variable tasks, dealing with potential health hazards. The problems of visual fatigue, back-ache, etc. are primarily to be found amongst full-time users engaged in repetitive operations. One reason for this is obvious; the longer

217

the exposure, the more likely any problems are to materialize. The other reason is not so obvious. Routine, repetitive tasks mean that the user is repeatedly using the same faculties and as a result they become fatigued. Thus the visual functions become fatigued and holding constrained postures for long periods leads to neck-ache, back-ache, etc. When the task is variable, its demands on its user are often variable and these problems are less likely to occur. Application of ergonomic principles to work-station design and to the design of the physical environment can alleviate these problems, but it would surely be more beneficial if we could avoid the creation of jobs with these characteristics. This chapter is an exploration of the possibilities and problems of job design in this area. Its theme is that prevention is better than cure.

GENERAL PROBLEMS OF REPETITIVE FULL-TIME VDU USAGE

Whilst the potential health hazards of full time VDU operation are of major concern, it is important to note that there are many other problems often (but not always) associated with these jobs. In Figure 2 these problems are listed as direct and indirect effects.

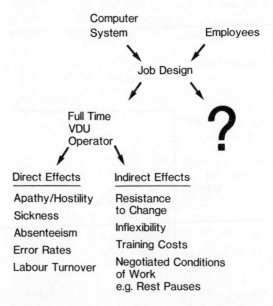

Figure 2. The effects of full-time VDU operation

When a job is highly repetitive and gives rise to visual fatigue, back-ache, etc. the workforce often displays apathy and even hostility to the needs of the business. There is often an increase in sickness and absenteeism and, where there are alternative forms of employment, labour turnover may increase. The combination of relatively meaningless work and poorly motivated staff frequently leads to high error rates.

The creation of this kind of work-climate leads to indirect effects. Poorly motivated staff who have a limited range of skills are a relatively inflexible resource who would be expensive to re-train and would probably be resistant to change. In addition, in an effort to protect their members, representatives may seek to negotiate many of the conditions of service (see Chapter 12). Rest-pauses is currently a popular topic in such negotiations and, inappropriately handled, can have the effect of creating a formal and rigid set of working conditions which threaten to bring the worst industrial relations practices of the shop-floor into the office, with consequent problems when swift adjustment is needed under pressures for change.

It seems that once we are committed to the creation of jobs characterized by full-time use of VDUs for repetitive work we are likely to be struggling with a complex array of problems, costly in terms of human well being and of the effectiveness of the organization. There are many ways in which these problems may be ameliorated, but the most constructive step would be to avoid getting into the situation. The purpose of this chapter is therefore to ask why jobs are created in this form (as Figure 2 illustrates) and to ask whether there are other job design opportunities.

CASE STUDIES IN THE CREATION OF VDU JOBS

To demonstrate that there are alternative forms of job design I intend to examine two cases where full-time VDU operation was a possibility but where there were also other possibilities. Following an examination of these cases we can use the lessons they have to offer to attempt a constructive strategy for job design.

Order Entry Systems

In this case a large organization had branches throughout the country which took orders from customers for a range of products. They organized the distribution of products and the invoicing of customers. The organization introduced an on-line order entry system which meant that it was possible to make a direct check on the availability of products while, for example, the customer was on the telephone. The system permitted on-line ordering and was also linked to order-picking and distribution from the warehouse, and credit control, invoicing, and accounts. A fuller description is given in Eason and Sell (1981).

No central directives were given on job design in relation to the system and the branches went their own ways in manning it. In some branches they created jobs with task or functional specialization, i.e. the presence of VDUs in the order reception unit was taken to mean that an operator should be trained to receive all the orders from customers and to process them into the computer. In another section an operator used the VDU for warehouse operations and in the accounts section an operator kept the accounts. In each section one person was given the specialized task of operating the VDU, and because of the volume and specific nature of the transactions, they became full-time operators with repetitive tasks.

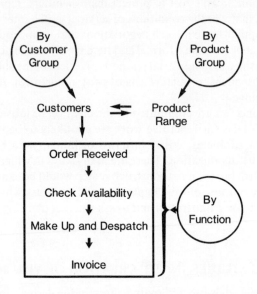

Figure 3. Job design options in an order entry system

Other branches took different routes. They took the opportunity of the new system to rethink their organization structure. In some cases instead of having sections organized by function they organized staff into product groups, so that orders for particular groups of products went to specialized members of staff who took the orders through the complete process, including the accounts. This had one effect which was important to the company; staff became knowledgeable, both technically and commercially, about the products with which they dealt. It also had the effect of giving each member of staff a range of tasks to undertake, some with and some without the VDU, and when they used the VDU it was for a variety of functions. Other branches organized staff into customer groupings so that a particular member of staff was the contact point for each regular customer and processed all work for them. This was beneficial for customer service and gave the staff a variety of tasks which were related to one another. In yet other branches a mixture of customer/product strategy was adopted because customers tended to deal with only a part of the range of products.

I do not wish to judge which of these manning methods is best, but merely to note that there are many different possibilities and that each one has different consequences for employees and for organizational effectiveness, e.g. quality of customer service.

Word Processing Systems

We have recently completed an investigation of job design in word processing systems as a part of our subscription research programme (Simpson, Eason, and Damodaran, 1980). We found once again that the jobs created showed considerable variation. One pattern was to provide a centralized service from a large shared-logic system run by a team of operators. In some of these systems there was a sub-division of function within the word processing system (option 4 in Figure 4). Some operators typed and corrected drafts, others operated printers, checked drafts, or allocated work to operators. This kind of system created full-time VDU operators. It achieved a high keystroke rate but often produced meaningless work for operators who had little contact with authors and rarely saw the same document through successive drafts. Authors tended to be very critical of the sensitivity of such systems to their special needs. Centralized systems of this kind do not have to be organized in this way. It is possible, for example, to retain an identity between an operator and a group of authors by allowing each operator to deal with the work from a User Section and to see this work through its complete cycle (option 3). This appears to produce more meaningful work for the operator, more variation in activity, and more sensitivity to author needs.

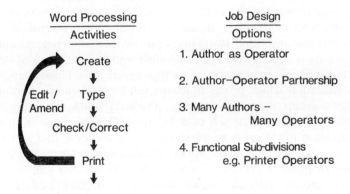

Figure 4. Job design options in word processing systems

In many word processing applications a stand-alone machine is placed in a User Section and exclusively serves authors in that section. In this case it is often used as a direct substitute for a typewriter. It is used when there is work appropriate for it and operators undertake a range of activities in the service of their authors, many unrelated to word processing (option 2). Again we may note that there are many different possibilities and that each one has consequences for the employee and for the organization. In passing, we may note that there is another possibility which at present, according to our investigations, is only a theoretical possibility. This is the case of the author using the word processor himself (option 1), an assumption of the 'automated' or 'paperless' office concept that still belongs to the future.

A STRATEGY FOR VDU JOB DESIGN

Having given examples of alternative forms of job design, in the remainder of the chapter I want to develop an outline strategy for constructive job design during systems design. The strategy considers three questions:

(1)   What are the theoretically possible ways of organizing jobs in relation to a specific system?
(2)   What criteria should be used to select between the possibilities?
(3)   Who should be involved in the job design process and how should they be involved?

Each of these questions is examined below.

Eliciting Alternatives

It is necessary to begin by explicitly asking what range of alternatives are hypothetically possible, because there is strong evidence that in most system design processes alternative formulations are not fully considered. It is unlikely that any single member of the design team will be given responsibility for job design. Jobs tend to become defined as a result of the actions of systems designers, user managers, and others in the pursuance of their normal work. For example, it is quite common for systems designers to assess the flow of work and to assign just sufficient terminals to cope with the peak load. This demands a high machine utilization which probably means full-time operators, because it is difficult for a group of operators to share a terminal and attain a high level of utilization. Consequently the job of a full-time VDU operator is created. If explicit attention is not given to job design the tendency is that jobs become functionally specialized, i.e. all the work of one function is collected together and handed to one or more operators for whom it becomes their job.

A very common implicit job design process goes something like the following. A computer system requires data to be input. What is more natural than to collect all the data together and give it to a dedicated data input operator? When the output emerges at the printer, other activities may be needed; sorting, stripping, collating, etc. Why not collect these activities together and give them to another operator? Two functionally specialized jobs are thus created. But other options are also possible, and our first requirement is to recognize the other options.

It is not sufficient merely to collect together a variety of unrelated tasks and call it a job. The most dominant theme in job design is the allocation of tasks to a job which display some variety but have a meaningful relationship one with another. Figure 5 lists the alternatives which we have found in many different organizations. It reflects particularly the options noted in the two cases discussed earlier in the chapter.

(1) **By function**
Single-function jobs
(2) **By product**
Multi-function jobs organized by product or service
(3) **By customer**
Multi-function jobs serving groups of
customers, suppliers, authors, etc.

Figure 5. Alternative forms of work organization

The most commonly cited alternative to functional specialization to be found in the job design literature, e.g. Davis and Taylor (1972), is by product or service type. It entails giving each job holder the opportunity to complete all the activities necessary to build a product or provide a service. Another possibility is to construct the job around all the duties required to serve an outside agent which gives or receives work (by customer, supplier, author, superior, client, etc.). In addition to these pure forms of organization, many mixtures may also be possible, i.e. some of the functions for some of the products.

It is remarkable how many different ways there are in which tasks can be organized into jobs. Unfortunately, in most organizations no effort is made to enumerate them, or any unusual ones that are suggested are immediately rejected on the grounds of cost or breaking the traditions of the organization. Job design is about choice, and we need options to choose between. Deeper examination of options that at first seem unrealistic is often rewarding. In our work we find that it is necessary to suspend judgement upon job design options until we have released the creative talents of organizational members in generating a lot of the possibilities. Then we can move on to the second phase, which is to expose them to a critical examination.

Evaluating Alternatives: Guidelines for Job Design

Selection of a form of job design will depend upon the evaluative criteria relevant in the organization. A list of the categories of criteria that might be considered relevant is given in Figure 6.

(1) Cost
(2) Productivity
(3) Technical
(4) Tradition
(5) Organizational effectiveness, e.g. flexibility
(6) Health and welfare
(7) Satisfaction and motivation

Figure 6. Categories of evaluation criteria

The first four items on this list are the factors which typically dominate the creation of jobs. The reasons, for example, for establishing a centralized word processing service are often the cost savings of reducing staff, achieving high throughput, and high machine utilization. If there has been a tradition of providing secretarial services from a centralized typing pool, it is also organizationally very much easier to implement this kind of system. But, as we have seen, if we adopt this path we may run into difficulties of employee health, motivation, and organizational effectiveness which, in themselves, are costly. Unfortunately, whilst in retrospect it is easy to see these are expensive, it is not easy to cost them in the evaluation which takes place during systems design. One important need we have is for accounting methods which can incorporate these wider and less measurable types of costs, a subject upon which Hopwood (1980) has made a useful contribution.

If we are able to bring these factors into the debate before jobs are fixed it might well be that jobs based on premises other than functional specialization will be seriously examined. We need, however, to demonstrate that these factors can be developed into a set of criteria which can be employed to evaluate alternative options. Organizational effectiveness can, for example, be considered in much broader terms than volume of output. It can consider future flexibility, quality of service to customers, the development of human resources, etc. Considerations of health and welfare in the context of VDU usage are, as we have suggested at the beginning of this chapter, ultimately linked to frequency of use and task type. The question of job criteria for positive employee satisfaction and motivation has been extensively researched by social scientists. The list of criteria employed by the Work Research Unit (Tynan, 1980) is given in Figure 7.

Tasks should as far as possible:
Form a coherent job
Make a significant and visible contribution
Provide variety of method
Allow 'feedback' in performance
Entail the use of discretion in carrying out the work
Carry attributable responsibility for outcomes and particularly control of work

Figure 7. Job design criteria (Tynan, 1980)

The important elements in any job from a job holder's perspective are that there should be a variety of tasks that relate together to form a coherent whole. The task-performer should have a degree of discretion in carrying out his tasks and should be able to obtain adequate feedback to assess and improve his performance. Finally, the complete job should make a significant contribution to the overall work system and the performer should carry responsibility for the outcomes of his work. It is widely recognized that, by following these principles, jobs may be designed which are worthwhile experiences for individuals and contribute to flexible and effective work systems.

It is important to note that the latter three types of evaluative criteria in Figure 6 tend to point in the same direction. The following list are criteria which relate these concepts of organizational effectiveness and employee health and well being directly to VDU jobs.

(1) *Include preparatory and consequential tasks in the jobs of VDU operators.* The flow of work usually means that the tasks that actually involve VDU usage are preceded and followed by tasks that relate to the transaction. If it is possible for the operator to engage in these tasks, the job becomes more variable, it achieves a degree of coherence because the transaction is followed through and the operator perceives the consequences of his own actions which provides feedback and reduces errors. The user is also no longer a full-time VDU operator or fixed in a constrained posture.

(2) *Include in the job both data input and tasks that involve using the data.* If a job consists entirely of data input, the data often lack inherent meaning, which can engender high error rates. The operator often perceives himself keying data into a vacuum and what happens to it is a mystery. If the operator can also perform tasks which mean that he is using the data held within the system, the meaning of the input task becomes clear and errors less likely.

(3) *Avoid loading a specific job with high volumes of data input.* Where there is a high volume of data input the effect on specific jobs should be minimized by

decentralizing its capture, e.g. capturing it at source, by using automatic recognition devices and by ensuring that data are only captured once, i.e. retaining standard data relating to customers, etc.

(4) *Avoid pacing by the computer or by the flow of work to the operator.* The chief enemy of discretion is pacing; an external control over the rate of work. It prevents the operator from setting his own targets, varying the rate at which he works, and having any control over what happens to him. In VDU based jobs there are two potential sources of pacing. The computer system by virtue of its response rate, by 'lock-outs' when there is no input in a given time, or, more explicitly, by logging user speed to provide statistics on operator performance can provide a variety of pressures on the operator.

The second source of pacing can be the flow of work to the operator which may have to be processed as it arrives. This is sometimes the case in VDU tasks which handle customer queries over the telephone or in person. In some cases the operator completing a task is instantly confronted by another task. If the user is to be given any responsibility or discretion, he needs some way of buffering the flow of work, of putting it into a temporary store, of selecting the sequence with which it will be handled and perhaps agreeing its allocation with colleagues. If users are aware of the total load and have the responsibility for getting through it, overall productivity is unlikely to be reduced.

(5) *Provide control over software sequences.* Many systems require the operator to follow a set sequence of activities to complete a task. The software that controls the interactive dialogue determines the order of the task. Not only does this remove any discretion the operator may desire over the way he handles the task but it means that the task has to be organized to fit the system. This can lead to ridiculous situations in which the operator has to write down details given in the wrong order or invent details he does not have available in order to proceed with the task. It can also mean that, if the operator is in conversation with a customer, he may have to take control of the conversation and elicit information in a set sequence rather than receive it in a customer-oriented way. This gives the impression of a system insensitive to customer needs and leads to fraught customer–operator relations. A system which allows the operator to control the software sequence provides the operator with the capability to adjust the task to local needs.

(6) *If the job has to be system-oriented, include related system tasks.* If it proves impossible to enrich the job with tasks related to the work flow and unrelated to system use it may be possible to enrich the job by giving the operator greater responsibility for some facet of the system. For example, the operator may become system adviser to enquirers, may take responsibility for the integrity of some of the files in the database, may undertake simple maintenance tasks on the terminal or work-station, or may act as liaison to system staff. Each of these activities is related to the main task, provides variation from it, enlarges the

operator's understanding of the function of the system, and provides a greater sense of contribution and visibility.

(7) *Avoid over-design.* The essence of good job design is to leave users with discretion. In the order entry case cited earlier different job structures emerged because the system permitted users to man it in different ways. One general criterion would therefore be to avoid system decisions, e.g. in hardware allocation or rigidity of software, which would prevent users adopting practices which suit their local needs.

Who makes the Choices? Mechanisms for Job Design

The final part of the strategy is to establish what mechanisms should be set up within system design to ensure that there is a consideration of job design issues. Who should make the choices between alternatives; who should place values on the benefits and costs of each scheme? Of one thing I am certain, it is not the right of an outside expert, especially not an ergonomist or psychologist, to determine the right or wrong solution for other people. They can point out the range of alternatives and the pros and cons of each alternative but the final judgement belongs to the people who must live with the consequences. This usually means the user management and staff who will work with the system after implementation. I would also exclude from this judgement the computer systems specialist who can point out the technical and cost implications of different solutions but has no right to assess what will be an effective or satisfying way for users to conduct their work.

We need, therefore, mechanisms whereby the people who will be most affected by job design decisions can play a role in the taking of these decisions. There are a number of specific strategies being advanced for this purpose. The best-known are those developed by Mumford and Henshall (1979), which either take the form of 'representational participation' (in which user representatives of all levels debate what is needed) or 'consensus participation' (in which attempts are made to involve all future users). In our work we have found that users need time to understand and come to terms with the change process and to learn to work on design options, and we have thus placed emphasis upon providing learning opportunities through pilot systems and trials (Eason and Sell, 1981) and, if possible, allowing systems to slowly evolve, giving everybody time to decide how best to exploit them. There is also the growing awareness of unions that it is in the interests of their members to be involved in the job design process. There are signs, for example, in new technology agreements that we are beginning to follow the pattern of Scandinavian countries in declaring the right of users to be involved in design decisions that will affect them.

It is doubtful that there is one best way of involving management and users in the job design side of systems design; the appropriate strategy depends upon the nature and size of the system, the normal pattern of industrial relations, the skills and expertise available, and many other factors. What is clear is that an 'expert behind closed doors' approach to job design is unlikely to be readily accepted.

SUMMARY AND CONCLUSIONS

The burden of this chapter has been that there is normally a missing link in the chain of systems design logic. If we proceed with primarily technical and economic criteria in mind we are likely to find that the system as implemented has human and social consequences. In particular I have pointed to the health hazards and other consequences of creating what appear to be cost-effective systems but which lead to operators using VDUs continuously for repetitive tasks. If we develop these solutions we run into all kinds of hidden costs, both in terms of human health and well being and of organizational flexibility and effectiveness. There may be real costs associated with sick leave, absenteeism, retraining, etc. and with steadily tightening procedures being negotiated between management and unions to deal with the effects of these situations.

The positive way out of this chain of consequences is to replace the missing link and to recognize that human and social criteria have a place in the systems design procedure; that we are designing socio-technical systems, not just technical systems, and that we need to pay attention to job design options. I have sketched a three stage strategy for examining job design which involves (1) a diagnostic phase in which the options are enumerated and (2) an evaluative phase in which options are examined for the degree to which they meet a wide array of criteria including good job design criteria. Finally (3), the strategy addresses the question of how these decisions should be taken and who should be taking them, pointing to the need to involve the people who will be most affected by them.

It will no doubt be argued that this strategy does not recognize the reality of many organizations. It may not, for example, fit the technical systems design goals or the preoccupation of senior management for precise cost-benefit evaluations based upon tangible gains. I would not argue that these are inappropriate criteria, but I would argue that they are insufficient in themselves; we have also to take into consideration broader organizational and human issues. The aim of this chapter is not to dictate job design solutions but to get job design explicitly onto the agenda in systems design and to have the human factor as one of the many sets of criteria upon which design decisions are based. It is pleasing to note that systems design theorists are beginning to shape their strategies to formally incorporate these issues. Gilb (1980), for example, has recently proposed a 'design by objectives' approach, in which each design stage is evaluated against an explicit array of technical, financial, and human goals. Until we can get human issues formally onto the agenda early in systems design we will continue to design systems on technical and financial criteria only to find ourselves surprised by the tangle of human and social consequences that result.

REFERENCES

Davis, L.E., and Taylor, J.C. (eds) (1972). *Design of Jobs*, Penguin, Harmondsworth.

Eason, K.D., and Sell, R.G. (1981). 'Case studies in job design for information processing tasks', in *Stress, Work Design and Productivity*, ed. by E.N. Corlett and J. Richardson, John Wiley, Chichester.

Gilb, T. (1980). 'A design by objectives approach', *Infotech Conference Proceedings on 'The Non-Expert User'*, Zurich, 22–24 September 1980, Pergamom-Infotech Ltd., Maidenhead.

Hopwood, A.G. (1980). 'Towards designing management accounting systems for the support of the new concepts of enterprise accountability', in *The Human Side of Information Processing*, ed. by N.B. Andersen, North-Holland, Amsterdam.

Mumford, E., and Henshall, D. (1979). *A Participative Approach to Computer System Design*, Associated Business Press, London.

Simpson, A., Eason, K.D., and Damodaran, L. (1981). 'Job design and training in word processor applications'. *HUSAT Subscription Research Report No. 2*, Loughborough University of Technology, Loughborough.

Tynan, O. (1980). 'Improving the quality of working life in the 1980s', *Work Research Unit Occasional Paper No 16*, Almack House, 26 King Street, London SW1Y 6RB.

Chapter 22

General Discussion

A psychologist from a telecommunications equipment supplier asked if there was any evidence to suggest that negative-ion deprivation leads to long-term cumulative effects. Hawkins thought not and added that it would be extremely difficult to set up a controlled experiment where one could be sure that it was only ions producing the long-term effect and not a multitude of other factors which could influence people.

Hawkins was asked if the negative-ion concentrations vary during the day and whether they are naturally lower during the night. He replied that indoors, in an air-conditioned building, there seemed to be no changes if the air-conditioning is running all the time. Air-conditioning very effectively buffers any changes which may occur outdoors so that there is a constant low level of ionization throughout the twenty-four hours. Outdoors there is quite a large circadian effect, which has been well documented in the literature. It seems to parallel solar intensity in a daily and seasonal rhythm, so that maximum levels of ionization would occur at midday on midsummer's day. Outside during the day ionization rises during the morning, reaches a peak at about lunchtime, and dips again during the night, although indoors it does not.

Another question criticized the experimental design suggesting that it would have been better to randomize the ionization over the period. Hawkins agreed and in subsequent experiments he had reversed the timing, with the ionization occurring in the first weeks and switched off later. He also randomized the ionization with weeks on and weeks off, but that produced some rather unexplainable, bizarre effects. The beneficial effects which occur when the ionization is on do not immediately disappear when it is switched off. There is a lag effect. In a trial, following a period of ionization the reduction in complaints was maintained for five weeks when no ionization occurred. A slow return to normal occurred during that period and it did return fully to the pre-ionization level.

An occupational hygienist from a commercial company pointed out that air-conditioning may effectively buffer the internal environment against the fluctuating ion levels outside but it had been his experience that it does not do

231

that against the climatic variables. Over a period of three months the external climate and therefore the internal climate will vary quite considerably. He wondered if Hawkins had taken any climatic data when measuring the ion levels and tried to relate that to the results. Hawkins replied that generally he had taken the measurements on clear, fine days but he had not made any correlations between indoor levels and climatic conditions.

Dr E. H. Burgess of Medical Advisory Service Department, Civil Service said that so many of our problems are due to lack of communication and many of the points raised by Sheila McKechnie were examples of this. For example, she said that no opportunity had been taken to build up evidence by carrying out the eye tests suggested. He was a member of the VET Committee and for two years in the Civil Service he had been conducting eye tests on a control group in the Metropolitan Police. The results had been negative and negative results do not usually have much interest or publicity. The unions had received this information.

Dr S. L. Gibbons, safety and security adviser, Imperial Chemical Industries Limited, said that his company's view was that a purchasing specification should refer to the safety of the equipment and ergonomics. Related factors are the responsibility of the managers who install it. They can seek advice from a whole range of personnel in ICI who are able to supply this.

S. McKechnie, of ASTMS, said that this leads to having one set of standards for the technical specifications and another set of standards, operated by a different group of people, for the ergonomic, 'user-acceptability'. Many companies buy equipment on the basis of the technical specifications alone, but the views of the health and safety, ergonomic, or medical departments should also be taken into account. In many companies these two groups never get together.

A human factors engineer from a telecommunications equipment supplier said that in his organization there was a very large group of about forty human factors engineers researching into new systems design and involved in various aspects of VDU design, environmental design, work design, and work organization. Over the past few months he had been using a questionnaire to investigate job characteristics such as autonomy and discretion, mentioned in the last chapter. The biggest problem had been in getting an agreement from the trades unions to do the research and he wondered if trades unions could be made to understand that he was really trying to do research that was for the operator's benefit.

S. McKechnie replied that the trade union movement was getting tired of academics and researchers carrying out research on them and going away to put the results in text books, never to be seen again. She was very interested in his project and would consider doing it jointly with him, as she would like to see shop stewards used as field researchers who could return to their members with the results and the applications of that research. For example, a group of bus drivers in Leeds decided that there is too much stress in their jobs and, with the help of a

WEA tutor, they organized a large research project. This research was not seen as a threat because of the way it had been set up and manner in which union representatives had been involved. His approach to that kind of research might lead nowhere because union representatives will not give company people access to confidential information which might be used against them. The only way that she could learn about the health of the whole workforce was to have access to the medical records. This would be refused to protect 'confidentiality'. Therefore, if unions cannot have access to their own members' records they are unlikely to let others have access, partly because of the uses to which medical records have been put. For instance, the occupational health records in many firms are given to insurance companies in order to fight claims. If he offered a different approach in his method of research, for example, its location and how it will be selected, then the unions might begin to co-operate with him. However, his question indicated a misunderstanding of the issues that would only produce a hostile reaction from union members.

The human factors engineer thought that S. McKechnie had made an incorrect assumption about the way he proposed to design the research. Research cannot be done by anybody, it needs a certain amount of training. On the whole, shop stewards do not have the sort of research skills that are necessary for this work. S. McKechnie replied that she found that most researchers have no communication skills and very little understanding of work organization. In the UK it seems that people and their capabilities are categorized, usually from birth, and that is where they tend to stay for the rest of their lives.

One questioner suggested that surveys of the type Leela Damodaran discussed were in most cases a thoroughly unparticipatory process which needed improvement. L. Damodaran replied that in her experience users have been able to provide very extensive information about all aspects of their day-to-day work. It is clear that analyses conducted by systems analysts focus upon the parts of the task which are to become computer-based while other aspects of the work, which may be regarded as of crucial importance to the user, simply do not get investigated. The real need is for a very extensive analysis of the user situation and very often the user is the best person to contribute to the systems analysis process. He occupies his workplace for seven or eight hours a day, knows all the operational details, the small daily problems and the seasonal variations and long-term problems. Consequently she would advocate involving the user from the beginning as he can be instrumental in setting up a method of assessing acceptability on a continuing basis.

S. McKechnie said that she had recently conducted a survey and some programmers seemed quite happy with the VDUs but were complaining about the computer print-out and the need to make them more acceptable to the user. Questionnaires tend not to pick up this sort of problem. The trouble that she and other union officials have is that they are twelve months behind the problems. A system needs to be devised to predict problems. However, if the system is totally dependent on the management it will be a waste of time. Damodaran said that there was no reason why print-out should not be well designed and well spaced. This also applied to the contents of the screens. Graphic designers should be involved in the design to avoid the disasters that many users have to cope with. There are one or two enlightened managements and client companies who are indeed trying to adopt a stance of looking five years hence. Instead of studying problems created by computer technology after the event they are trying to design systems from the outset which better match human capabilities. So there is a willingness on the part of some companies, it may not be for people-centred or economic reasons, but, whatever the motivation, there is concern to avoid problems.

It is inevitable that one receives retrospective reports of problems, and for this reason we need mechanisms for handling them on a more regular basis. One of the reasons she advocated regular audits of acceptability was so that problems are looked at before they become chronic or so serious that the system grinds to a halt. In addition, people are needed whose responsibility it is to do something about those problems. Here we come full cycle again. We have sound ergonomic, human factor, and health and safety policies but who is ultimately going to take the responsibility? On the basis of the contents of Tom Gilb's chapter it appears that responsibility lies with senior management. Only when corporate decision-making reflects a real concern for human factors can we expect the kind of changes that we would hope to see.

T. Gilb said that a system he used called 'evolutionary delivery' is a way of getting a user-acceptability test and feedback that has a realistic chance of changing the system to make it more acceptable. The evolutionary delivery of the system means that it is not installed in one phase and then assessed for user acceptability, nor is it installed in three or four phases. A global design with clear objectives had to be performed first. Then, instead of trying to install the largest section, the smallest increment of system is found which can be installed quickly and which will be highly valued by the user. One can deliver only 1–5% of the system at a step, perhaps to hundreds of users at once. This gives only a small evolutionary change from the old system. Using this method the old environment might be totally replaced within five years. The point of the system is that with a small increment the cause and effect can be measured more accurately; the increment which will cause satisfaction or dissatisfaction can be seen. If there is a positive reaction the system should be built upon. With a negative reaction it is practical economics to totally discard that 1% step. When setting the system up he explains that it is only an experiment but whatever the result, he has succeeded. If it is a success he leaves it, if it fails he has learned what was wrong and can then hope to install it correctly. Here it becomes practical for the system's designers to listen to the users instead of using the argument that £5 million had been spent and the users will have to tolerate the results, however unpleasant. This system not only gives the user a feedback opportunity but is also economically sound. In the *IBM Systems Journal*, No. 4 (1980) Harlan Mills discusses a project of over 200 person-years which was delivered in forty-five increments over four years, each of them on time and under budget. From the user's point of view it is a powerful and realistic, rather than academic, way of letting him view the system, but little discussion of the technique occurs.

L. Damodaran added that if that method is applied one must make sure that the first increment given to the user is very likely to be successful and acceptable, because one may not get a second chance. It is essential to warn users that it is an experiment but often the user sees it differently, and says 'They are playing games again. They don't know what they are doing'. They have to have something useful and valuable from the outset to build upon. From the ergonomic point of view there are a variety of techniques for working things out with users, which include pilot schemes, experiments, and actually testing the system. This operates in much the same way as when buying any other consumer products. When buying a washing machine, a television, or a car one usually has a practice run or its equivalent. This should be well structured so that the user can participate before one is committed too far along the road in the purchase of hardware, software, etc.

A national official of a trade union pointed out that when choosing a washing machine or car one reads about it in the appropriate magazines before purchase. These audits of user acceptability seem to be saying that a company has to buy 5000 VDUs and then ask the users whether they actually like them and whether they function well on the task for which they have been provided. He was concerned to find some way for unions and management to find out what acceptability had already been found by other users; in some confidential coded form if need be. Is it necessary to buy a machine before we can find out whether it falls to pieces in two years or whether the characters rub off the keys? Is there no centre like the Consumer Association and *Which?* magazine where we can get this information? Could we not set up an information exchange?

Damodaran said that a regular feature of conferences is someone stating the real need for a co-ordinating body to provide this kind of expertise. Regrettably, the government and research councils have not seen the need to fulfil that requirement. It is up to people such as the delegates to make that demand regularly, loudly, and as publicly as possible. There are ways of finding out about a technology and its impact other than by buying it and using it. There are many ways to ensure that one has a user-friendly system without going near the computer technology. Using an overhead projector one can experiment with different kinds of format, coding, and ways of presenting information to the user. The user can design what he wants to see on the screen and can make many decisions that are quite separate from the purchase of hardware or software. There are a variety of ways of conducting trials and simulations. These need not be full-powered computer models; cardboard boxes can be used as 'mock-ups' in work-station design and equipment positioning trials. Basic questions like that can be answered without having to invest in the technology.

T. Gilb gave an example of another mechanism which could be included. A New York electricity company has a system of 1800 terminal VDU operators who are mainly preparing electricity bills. The system is about nine years old with over 3000 people trained to use it and has a 30% turnover rate each year, not untypical for the United States. He discovered some problems in the system which the operators were perfectly well aware of but simply tolerated. There was no feedback mechanism from the terminal operators to the people who could authorize changes in the system. When the system was first originated the designer envisaged a small group of people working together who could communicate directly. As the system grew there was a need for a specifically designed feedback mechanism, but this had never occurred. He had discussed ways of counteracting this. For example, he could have built in the capability, at the supervisory level, of changing the HELP and error messages so that they explained things the way the operators liked them to. The designer himself does not need to design these messages in advance but needs to put in a facility to allow users to redesign the messages themselves. Changes can be then made without having to go back through the hierarchy to the designer. Users can, to a certain degree, redesign systems themselves.

Oliver Tynan, Director of the Work Research Unit, commented that the list (Figure 7, Chapter 21) had been endorsed, not just in principle but in detail, by the Employment Policy and Organization Committee of the TUC, the General Council of the TUC, and was passed by the Trades Union Congress in 1980. It was also endorsed by the Employment Committee of the CBI and is government policy. The list has therefore rather more strength than perhaps at first sight appears, and he could show many cases where it increases efficiency. When asked by a trade union official why more companies didn't therefore adopt the approach Oliver Tynan suggested that one of the problems was that the participatory option, whether formalized or consensus, is full of risk. When consulting others there is always a risk that the decisions will have to be altered. One starts at A and wishes to get to B, but one may end up at a different B or have to move to C. In engineering, millions of pounds are being spent on systems and machinery. The boards, and people spending the money, are going to require dates and return on capital, things which are difficult to be absolutely sure about when participatory decisions are made.

A delegate said that his company was very small and they had varied employees' jobs and they were happy to do this, but problems arise, particularly if a job is partly clerical and partly despatch, for example. Could one get a person physically able, or willing to sit in front of a VDU one minute and the next put on an overall and carry half a ton of product?

K. D. Eason pointed out that this illustrated why he would not offer any specific recommendations to any firm without knowing something about them.

The illustration given was exactly the sort of thing that clerical users will give. They will say that that type of solution cannot be implemented because there are two geographical locations and that the two activities are incompatible. On the other hand, they may debate whether it would be possible to organize such a system. He would ask them to consider it but he would certainly not tell them to do it one way or the other.

Oliver Tynan said that the list that Ken Eason used was of provable generalities which must always be applied in the particular. The only way of getting true evidence was by asking for the worker's opinions, not by making assumptions about their jobs. One man in a pork-pie factory had spent thirty-seven years boning pigs' heads and loved the job. There is no substitute for asking people about their views of their jobs. It is also important to ask people for their opinions so that their talents are not lost. For example, a large ethylene plant incorporated a fail-safe system to turn the power off which required two valves 185 feet apart to be turned at the same time, and two men had to be employed to do this. If the operators had been consulted the design engineers would never have planned it that way. Unfortunately, designers often do not talk to people that operate plants.

K. D. Eason said that when considering different participation strategies for designing jobs we must be concerned with giving support to the people who are, perhaps for the first time, being encouraged to participate in creating the circumstances in which they are going to work. This is a good idea but it can be quite frightening. We have seen examples of systems designers who, with the best will in the world, approach users with a flow chart of a proposed system and ask for opinions. They go away disappointed with the response and do not do it again. We have to do a lot in terms of giving something concrete to people to respond to. It may involve taking chairs for them to see, using cardboard boxes in the work-stations, or showing them examples of what the dialogues might be like. They must definitely be shown more than one example, otherwise they will think that is the way it is going to be. This is what job design is about—helping people to work through such processes.

L. Damodaran said that when helping people to conduct analyses and research it is important to realize that there is a wide range of individuals in most organizations who already have the relevant experience. In the case previously discussed, one of those groups of individuals would indeed be the shop stewards. Although people needed interviewing skills, skills in which shop stewards would require training, they already have the trust of the people they represent and an understanding that they would respect the data being obtained. There is a real need to reconsider the role of research and who should be conducting it; to move away from the very restrictive model of only academics or consultants with the appropriate training, guidance, and support conducting research. One of the roles increasingly adopted is to operate as a catalyst to help people integrate existing skills; to provide a new way of looking at problems and a set of techniques and procedures to be used by quite a wide range of personnel within an organization.

Name Index

239

Subject Index